THE ENVIRONMENT GOES TO MARKET

The Implementation of Economic Incentives for Pollution Control

National Academy of Public Administration
July 1994

PANEL MEMBERS

National Academy of Public Administration

Richard A. Wegman, Chair
Partner, Garvey, Schubert & Barer

Alvin Alm
Senior Vice President,
Science Applications International Corporation

Douglas Costle
Former Administrator, U.S. Environmental Protection Agency

J. Clarence Davies
Director, Center Risk Management, Resources for the Future

William Drayton
Chairman, Ashoka Society

Anthony Earl
Partner, Quarles and Brady

Frank B. Friedman
Partner, McClintock, Weston, Benshoof, Rochefort,
Rubalcava & MacCuish

Robert Hahn
Resident Scholar, American Enterprise Institute

David Hawkins
Senior Attorney, Natural Resources Defense Council

James M. Lents
Executive Director,
South Coast Air Quality Management District

Edward Sanders
President, Sanders International, Inc.

Richard Stewart
Professor of Law, New York University School of Law

Victoria Tschkinkel
Senior Consultant, Landers & Parsons

PROJECT STAFF

John Palmisano
Project Director
Carole Neves
Deputy Project Director
William Foskett
Research Associate
Konstantin Gofman
Research Associate
Brent Haddad
Research Associate
DeWitt John
Research Associate
Kseniya Lvovsky
Research Associate
Elizabeth Epstein
Research Assistant
Julia K. Oster
Research Assistant
Anne Edwards
Manuscript Coordinator
Whitney Watriss
Editor

REVIEWERS

Jack Broadbent
Brian Cooke
Robert Friedman
Barry Korb

CONTENTS

FOREWORD	ix
PREFACE	xi
INTRODUCTION	1
CHAPTER 1: TWO APPROACHES TO ENVIRONMENTAL CONTROL	5
CHAPTER 2: MARKETABLE PERMITS AND POLLUTION CHARGES: TWO CASE STUDIES	21
CASE STUDY: South Coast Air Quality Management District and the Marketable Permits Program	24
CASE STUDY: The Russian Pollution Charge System	72
CHAPTER 3: CHARGES FOR SERVICES AND DEPOSIT-REFUND SYSTEMS	113
CASE STUDY: Economic Incentives for Managing Waste in King County, Washington	115
CASE STUDY: Michigan's Mandatory Deposit Legislation	138
CHAPTER 4: APPLICATIONS OF ECONOMIC INSTRUMENTS	161
GLOSSARY OF ACRONYMS AND ABBREVIATIONS	187
SELECTED BIBLIOGRAPHY	189
PROJECT DIRECTORS	211
PANEL MEMBERS	213

ILLUSTRATIONS AND FIGURES
Chapter 2
2-1. South Coast Air Quality Management District Jurisdictional Boundaries ... 25

viii   *The Environment Goes to Market*

2-2. South Coast Air Quality Management District Organizational Chart — 37

2-3. Organizational Structure of Environmental Management in the Territories of the Russian Federation — 91

Chapter 3

3-1. King County Solid Waste Operations — 126

3-2. How the Michigan Deposit-Refund System Works — 145

TABLES

Chapter 2

2-1. National Ambient Air Quality Standards — 28

2-2. Three-Year Revenue Forecast by Major Account — 41

2-3. Three-Year Program Forecast by Category — 43

2-4. Offset Trading in the South Coast Air Quality Management District, 1985-92 — 50-52

2-5. Comparison of RECLAIM and New Source Review — 62

2-6. Confirmed Standards for Fees for Pollution of an Air Basin — 80

2-7. Trends in the Use of Fund Revenues in the City of Nizhnii Novgorod for the Protection of Nature — 83

2-8. Charge Rates for Emissions, 1991 — 97

Chapter 3

3-1. King County Solid Waste Disposal Fees, 1981-92 — 120

3-2. Residential Solid Waste and Recycling Collection Rates — 121

3-3. King County Division of Solid Waste — 124

3-4. King County Mixed Municipal Solid Waste Figures and Projections — 131

3-5. Costs Borne by Soft Drink Bottlers, 1988 — 151

FOREWORD

The whole way of thinking about the role of government and the management of public organizations is changing. Perhaps the most visible manifestation of change is the "reinventing government" movement, which encourages public-private cooperation and the use of market techniques. Even before, public agencies had begun to focus on approaches that placed greater reliance on the private sector. In the environmental arena, economic incentive programs such as fee and deposit-refund systems, which embody market principles, were used to complement or substitute for traditional command-and-control-style regulations.

The National Academy of Public Administration (NAPA), whose charter is to improve the efficiency and effectiveness of government at all levels, is encouraged by the new regulatory approaches. It has noted, however, that the practical aspects of their administration have not been well-explored, although much of the success of incentive programs depends on how well public entities design and implement market-based approaches. The Academy attaches great importance to the development of a strong understanding of the practical aspects of administering market-based programs and government's new role so as to facilitate adoption of these approaches and improve their performance.

This report, prepared by NAPA, focuses on the administration of market-based programs. Specifically, it describes and analyzes a number of implementation issues as evidenced in four case studies: the air trading program of the South Coast Air Quality Management District in southern California; the national system of pollution charges in Russia; the King County, Washington, solid waste recycling program; and the deposit-refund program in Michigan. These case studies point out the many factors that need to be considered before adopting market-based approaches, such as the relationship between the market-based programs and existing regulatory arrangements. The case studies also point to the important roles of government and the private sector in implementing market-based programs, particularly with respect to creating the initial conditions needed to carry them out, designing the programs, and evaluating them to determine whether the desired results are being obtained and whether the programs need to be modified.

The case studies also raise important issues related to the administrative capacities needed to manage market-based initiatives and the administrative restructuring that must accompany their adoption. For example, how do the public organizations responsible for implementing incentive programs develop and sustain the entrepreneurial capacity to manage resources? How do regulators collect and process information to measure the performance of incentive programs? How do the differing values and operating tenets of the public and private sectors affect the performance of market-based programs?

These and the other questions pursued in this report could not have been addressed without the able research skills and capabilities of project staff. I wish to thank them. The Academy is also deeply grateful to the panel members and reviewers, who shared their knowledge of and experience with regulatory mechanisms. The case studies benefitted greatly from their wide range of perspectives and, in many instances, demanding and constructive comments on the various parts of the manuscript.

Finally, this report would not be complete without an expression of gratitude to the Joyce Foundation for its financial support.

R. Scott Fosler
President, National Academy of
Public Administration

PREFACE

During the 1970s, there were a number of proposals aimed at making greater use of economic incentives in the implementation of new programs regulating solid-waste disposal, and air and water quality. Some proposed controlling sulfur dioxide (SO_2) pollution by levying a tax on industrial and utility SO_2 emissions. Others called for a per-pound fee on all components of the solid-waste stream. Still others suggested that imposing effluent charges for industrial wastewater discharges would simultaneously reduce pollution and generate funds to pay for new treatment.

Some viewed these proposals as licenses to pollute, fearing they would make compliance difficult to monitor and lead to widespread cheating. In addition, some felt such programs had the potential to undermine the authority of regulatory agencies, which traditionally make basic decisions about what reductions to achieve and how to achieve them. Together, these reservations were a major reason why the "marketplace" proposals did not gain widespread acceptance in the early years of most federal and state environmental programs.

Today, much of that thinking has changed. Policy-makers have become much more open to the use of economic instruments to achieve environmental goals. Local, state and federal governments increasingly have begun to incorporate fee systems, charges, subsidies, deposits and trading mechanisms—almost everything but direct taxes—into their environmental programs. One of the most significant developments is the trading mechanism for acid rain control that Congress included in the 1990 Clean Air Act Amendments. A similar trend has occurred overseas, where use of tax mechanisms is much more acceptable than it is here.

This study by the National Academy of Public Administration (NAPA) looks at four programs, one at the county level, one at the state level, one federal program, and one overseas effort. They are: the air trading program of the South Coast Air Quality Management District (SCAQMD) in southern California; the national system of pollution charges in Russia; the King County, Washington, solid waste recycling program; and the deposit-refund program in Michigan. SCAQMD's program uses pollutant trading to provide greater flexibility in meeting pre-set pollution goals and standards. The Russian program involves direct taxes imposed on industrial activity. The other two programs involve fee structures aimed at activities of the general public.

Although the results vary, all four programs show promise. In three of them, pollution levels have dropped since the incentive endeavors were implemented. (Southern California's was implemented only a few months ago; its results are not yet available.) In all four cases, the officials responsible for implementation are strongly committed to using marketplace incentives to achieve programmatic goals. King County and southern California officials are actively exploring how to expand their programs.

Unfortunately, it was not possible to assess explicitly the benefits of using economic incentives in the cases NAPA studied; comparing those benefits with those of traditional command-and-control approaches would have been very useful. Such an assessment could help regulatory officials make mid-course adjustments, increase or decrease staffing levels and identify potential areas for expansion. A major finding of the study is that each program administrator needs to do a much better job of evaluation—a mechanism which needs to be built into each program at the time it is adopted.

Several common threads emerged from the case studies. In each of the jurisdictions, public support for the economic incentive program appears to be quite strong. Consumers and business alike seem to welcome the opportunity to participate more fully in safeguarding the environment, while simultaneously furthering their own self-interests. Economic incentive mechanisms also have significant impacts on agency cultures. Program administrators are generally strongly committed to programmatic goals and, as a result, tend to be very solicitous of public opinion. Many of these programs are relatively new, of course, and it remains to be seen whether these attitudes will persist. For the present, however, agency officials manifest strong desires to see their programs succeed, attitudes that are likely to be self-fulfilling.

Each of the four programs appears capable of achieving a significant reduction in the level of government involvement in personal and private business decision-making—the major goal of any economic incentive program. Of the four case studies, Michigan's bottle bill is perhaps the closest to self-enforcing, while the southern California program probably has the farthest to go in that regard. Over time, however, all of these programs offer the hope—and the promise—that regulators will be freed to concentrate primarily on what they do best in the environmental field: set goals and targets for reducing pollution levels. Decisions about how to meet the goals and targets will be left largely to individual consumers and businesses themselves.

A note of caution is in order. Environmental policy-makers should not view economic incentives as a panacea. These mechanisms are not appropriate for handling all environmental problems, perhaps not even for addressing a majority of them. Incentives generally need to be designed to fit within the larger regulatory system, to be employed judiciously as underlying problems dictate. However, where the proper conditions exist, it is highly likely that pollution reduction can be achieved more efficiently and at lower overall cost than would be possible with a command-and-control system.

The chart below, "Objectives and Applicability of Incentives," summarizes the characteristics of the economic incentive mechanisms considered in this study. The last two columns suggest the circumstances in which they are most likely and least likely to be effective.

On behalf of the panel, I want to thank the project staff, so ably headed by John Palmisano and Carole Neves, for its exceptional effort in preparing this report. An interdisciplinary group of experts, drawn from regulated businesses, environmental groups, administrative agencies, legislative organizations and academia, drafted the case studies. The panel very much appreciates its first-rate efforts. Thanks also go to the National Academy of Public Administration and President R. Scott Fosler for giving the panel the opportunity to participate in this important and challenging endeavor. The panel is also grateful to the Joyce Foundation for the generous financial contribution that made this study possible.

All the members of the panel hope this report will encourage greater use of evaluation techniques for marketplace incentives and will help policy-makers identify the conditions in which such mechanisms are most likely to be useful. If the report achieves even a part of this goal, the effort will have been very worthwhile.

Richard A. Wegman
Panel Chair

OBJECTIVES AND APPLICABILITY OF INCENTIVES

Objectives	Mechanism	Outcome	Effectiveness	Administrative Complexity	Analytical Requirements	Effective When	Ineffective When
1. To be a substitute for or supplement to command-and-control regulations	Pollution taxes (Russia, King County)	Known costs, unknown results	Unknown	Potentially simple (but probably complex because of political concerns)	Extremely complex	Pollution is homogeneous (e.g., SO_x/NO_x nutrients) and extends over a broad area	Impacts are heavily localized (unless taxes consider harm done to human health and the environment)
2. To create closed-loop systems	Deposit-refund systems (Michigan)	Generally effective	High	Low	Low	Disposal of item adds to litter, solid waste, or creates other environmental problems	Environmental problems are caused by effluents or emissions
3. To encourage experimental investments (usually in conjunction with regulations) to reduce costs	Marketable permits, bubbles, etc. (South Coast Air Quality Management District)	Determined by legislative fiat	Potentially high	Very high	Low	Pollution is homogeneous	Impacts are heavily localized

INTRODUCTION

Command-and-control modes of regulation have dominated environmental protection programs since the inception of environmental regulation. However, a significant trend in regulation has emerged during the last 20 years—the use of economic incentives. In the area of the environment, a vast literature has emerged on the evolution of environmental regulation, including the use of economic incentives.

This literature rarely touches upon the implementation of economic incentives for environmental protection. What information exists comes from five sources—economists, attorneys, policy analysts, regulatory agencies, and the general press. Economists tend to produce advocacy pieces laden with theory, cost-benefit analyses and associated firm-level economic analyses of cost savings. Articles by attorneys and policy analysts generally focus on the nature of property rights, enforceability, and legal or regulatory fixes that are needed to promote one tool (taxes, for example) over another (such as marketable permits). Regulatory agencies themselves have produced a great amount of material supporting the status quo, floating trial balloons for proposed regulatory changes and advocating positions adopted by the agency. Articles in the lay press on the use of economic instruments are often planted by particular stakeholder groups and are thus biased, or they take a "man-bites-dog" approach to the novelty of designing an economic incentive-based environmental program. During the past two decades, there has been almost no serious assessment of the implementation issues associated with the use of economic incentive-based reforms such as offsets.

Interest in economic tools has grown recently as the limits of the traditional command-and-control approach to environmental protection have become evident. Legislatures and regulators adopted command-and-control modes of regulation during the first phase of environmental regulation, and these programs have produced significant successes during the last 20 years. However, they are often administratively complex and costly, and over the long run, they may stifle technological innovation. For example, over the life cycles of traditional regulatory programs (such as the U.S. Environmental Protection Agency's program to control ambient air quality), regulators find they have to fine-tune the instruments to produce more-cost-effective outcomes and reduce the administrative burden on complying firms.

Despite the enormous literature endorsing the use of economic incentive-based programs, not until the early 1970s did legislators and regulators begin to look to reform programs that use the positive features of the marketplace and other incentive-based approaches. Now they do so routinely. In recent years, Russia (and its predecessor, the USSR), Germany, France and the United States have increasingly been adopting this approach. They have been joined by a broad spectrum of environmental advocates. Russia's entire environmental protection system is now based on the use of discharge fees (taxes) to influence pollution control practices related to air, water, and solid waste. In 1990, the U.S. Congress instituted a pollution control policy whose foundation is market-based concepts. One result, the sulfur dioxide (SO_2) allowance trading concept, is generally accepted as a promising program producing good environmental outcomes at less cost to society.[1]

This report looks at four distinct economic incentive-based programs for environmental management—the air credit trading program in the Los Angeles metropolitan area, the national pollution charge system of Russia, the recycling initiative of King County, Washington, and the deposit-refund system of Michigan. The four programs, which address different environmental problems and geographical and political circumstances, involve different tools. The air credit trading program, which has had an impact in the Los Angeles area, is metamorphosizing into a full-blown marketable permit system for managing emissions from both new and existing stationary sources of nitrogen oxide (NO_x) and hydrocarbon emissions. The Russian pollution charge program is the most ambitious use of a national economic incentive-based pollution control program in the world. The King County recycling initiative is an example of a program that uses variable fees and a series of related tools such as education to promote alternative uses of discarded products by individuals and industry. The Michigan deposit-refund program imposes a refundable tax on beverage containers to encourage individuals and groups to collect and return them.

The information in this report comes from case studies of the four programs. The case studies involved extensive interviews with various stakeholders and reviews of the literature on the programs. The objective was to identify implementation issues that cut across the programs and that would be of interest and use to designers and implementers of environmental reform efforts. The issues relate both to the start-up and implementation of the four economic incentive-based reform programs, as well as to their fine-tuning down the road. The report centers on such points as the institutional, organizational, informational and administrative requirements for successful operation of the programs; their acceptability; conflict resolution; their compatibility with other regulatory schemes; and program evaluation. Given this focus, the report does not present a great deal of statistical material on costs, economic benefits, firm-level activities and the like.

Chapter 1 provides a general overview of the characteristics and use of economic instruments for environmental protection. Chapter 2 addresses marketable permits and pollution charges, the first as implemented by the South Coast Air Quality Management District in the Los Angeles area and the second by Russia (and the former Soviet Union). Before looking specifically at

---

[1] See also Robert W. Hahn and Carole A. May, "The Behavior of the Allowance Market: Theory and Evidence," *The Electricity Journal* (American Enterprise Institute) (forthcoming).

those two case studies, chapter 2 discusses briefly some principles and uses of these types of economic programs.

Charges for services and deposit-refund programs, as implemented by King County in its recycling program in the first instance and by Michigan in the second, are addressed in chapter 3, which likewise opens with some general comments on these types of programs.

The report concludes with chapter 4, which presents the main findings and conclusions from the case studies. The chapter looks first at each economic instrument in terms of the environmental problems for which it is appropriate and then at how to design each instrument so it will be a success. The report then addresses a critical activity that was largely absent from the design of the four programs studied—systematic evaluation. The focus is on the roles systematic evaluation can play in the adoption and implementation of economic instruments. Suggestions on how to design and structure an evaluation system as an integral part of the economic incentive program are presented.

1
TWO APPROACHES TO ENVIRONMENTAL CONTROL

The use of economic instruments for environmental protection is based on the concept that private managers necessarily have more complete information on their operations than regulators do. If regulators offer these managers appropriate incentives, industry can and will devise more effective and less expensive solutions to pollution problems. At a theoretical level, economic instruments offer the potential for increased regulatory flexibility, lower pollution abatement costs, and accelerated progress toward environmental goals. Because of these properties and the results of limited experiments, both regulators and private managers have expressed considerable interest in the implementation of economic instruments at the national, regional and international levels.

Economic incentives encompass a range of specific instruments, including:

- Monetary charges, which involve a fee for each unit of pollution a firm produces.

- Subsidies, which provide financial assistance that serves as an incentive to polluters to alter their behavior so as to meet the goals of the regulator.

- Deposit-refund systems, which focus on minimizing waste or preventing the production or improper disposal of pollutants.

- Financial enforcement incentives, which mandate compliance by requiring a deposit that is returned upon compliance or by fining a polluter that fails to comply with a standard.

- Market creation, which sets up markets within which firms can buy or sell environmental credits for actual or potential quantities of pollution.

This chapter looks first at the characteristics of traditional command-and-control regulatory approaches and at the limitations to this approach that have led to increased interest in economic incentives for environmental protection. As background to the four case studies of specific economic incentive programs presented in the next two chapters, it then discusses the nature and application of economic incentive programs in general and examines four approaches—tradable emission permits, pollution charges used to support environmental funds, a portfolio of incentives to encourage recycling, and deposit-refund programs.

TRADITIONAL REGULATORY APPROACHES

Governments use a variety of direct regulatory approaches to protect the environment, such as technology requirements, permits for facility emissions and discharges, limitations on chemical risks that range from total bans or minimal tolerance, and criminal penalties for violations. These approaches are commonly referred to as command-and-control policies: regulators "command" polluting firms to "control" their pollution by specifying what they should do and how they should do it. (See box 1-1 on the rationale for government intervention.) That is, typical command-and-control programs involve direct regulation of the quantity of pollution individual sources are allowed to emit or specify the pollution control technology they must use.

BOX 1-1
THE RATIONALE FOR GOVERNMENT INTERVENTION

Government intervention is needed to achieve environmental goals because private market solutions are less than optimal. In most cases, private markets for environmental resources do not exist. Private decisions about the uses of environmental resources are therefore often not based on the full social costs of those uses. Without government intervention, the private sector will engage in too many activities that consume too many resources and create environmental costs, without at the same time counterbalancing them with adequate investment in environmental protection. However, elimination of all externalities is not practical or desirable. Finding the balance between too much pollution and excessive control of pollution is the job of both legislators and regulators.

Absent market failure, environmental degradation may not by itself be sufficient to justify government action. After all the externalities have been internalized to decision-makers, pollution may continue to exist. Cost-beneficial policies must reflect the trade-offs between environmental protection and other societal goals. Therefore, the first step in evaluating any policy proposal is to determine whether and why the private market has failed and what adjustments are required to redress environmental market failures. Government intervention should be designed to encourage the socially optimal amount of environmental protection using a balance between "carrots" and "sticks." Government intervention is also needed to integrate command-and-control programs with market-based initiatives. In addition, policies need to be crafted to ensure that policy targets are hit and negative unintended results minimized. Clearly, regulating complex economic systems is difficult; however by divorcing ends from means, more cost-effective solutions seem to be achievable.

The direct regulatory approach is predicated on the concept that regulators can determine technically and economically feasible actions to control specific pollutants and that, by using these technologies, there is a basis for enforcement. As a result, regulation in each medium (air, water, and land disposal) attempts to assure that media/industry-specific design standards will be developed. In some cases, uniform design-based technology is required nationally. For example, selective catalytic reduction is required for certain new sources of nitrogen oxide (NO_x) emissions in the United States. Many of the environmental statutes mandate that the U.S. Environmental Protection Agency (EPA) translate regulatory goals into ambient standards, that is, levels of environmental quality necessary to meet the goals. Subsequently EPA specifies the controls necessary to meet those standards for each industry, a process often guided by statutory language that requires so-called best technology. In other cases, state or local governments must implement programs that achieve the national environmental quality standards. These governments must choose between a design standard approach or other approaches to regulation such as performance-based ones. A performance standard is based on a single target such as a discharge limit for a pollutant of, for example, 10 parts per million or 50 pounds per hour. Firms are free to meet the goal in whatever manner makes most sense to them. Firms must still meet appropriate record-keeping and monitoring requirements, but under a performance standard approach they are left to develop the best pollution control system for their affected plant or collection of plants.

Standards may differ between new and existing facilities. Inasmuch as environmental standards affect industrial survival, controls over existing sources are often more lax than those for new sources. New sources have the opportunity, relatively easily, to build in the best control technologies, whereas existing sources may find installing today's best technology requires an expensive retrofit. Under EPA's air program, for example, the pollution control technologies required for new sources may be design standards, while the retrofit requirements for existing sources tend to be more performance standard-oriented. The assumption underlying requirements for the strictest control levels on new plants is that new sources can more easily incorporate pollution control technologies than can existing sources and that economic growth will result in the rapid replacement of older, dirtier facilities. Slower than expected growth in the manufacturing and electric power sectors since the mid-seventies has, however, belied that assumption. In addition, industry's desire to protect its investments in old capital stock has kept those facilities in operation longer than expected.

Current command-and-control policies have scored notable environmental successes. Many water bodies have much less organic pollution compared with 30 years ago. The levels of hydrocarbons, sulfur dioxide, ozone, and particulate matter in various urban areas are much improved over what they were in the late sixties. It has become increasingly clear, however, that reliance on the command-and-control approach to environmental regulation will not, by itself, allow EPA to achieve either its mission or long-established environmental goals. A number of seemingly intractable problems persist. The current system has, for example, been unable to achieve many mandatory ambient air and water quality standards, and deadlines established in the authorizing legislation are continually being moved further and further into the future. Regulators have not achieved the ambient air quality standards for ozone and carbon monoxide in dozens of urban areas, and many bodies of water still do not approach their mandated quality. Whereas in the past the regulatory focus had been mainly on controlling pollution from large, industrial sources, now environmental mandates from Congress and regulators mean that a diverse range of regulations may be imposed on small and large sources, consumer products, and previously unregulated activities. Prior regulatory approaches that might have worked with steel plants may not work when regulating the hydrocarbon content of ballpoint pens, architectural coatings, or acid deposition. Hence the need for greater diversification and the use of new regulatory tools has arisen.

In addition, for many problems, traditional command-and-control approaches may not be practical at an acceptable cost. Examples include non-point source pollution from pesticide and fertilizer run-off into surface and ground water; solid waste disposal; stratospheric ozone depletion; and the buildup of pesticides and toxic materials in land, water, and air. Combustion by-products and other industrial chemicals affecting the global climate may be the ultimate pervasive environmental challenge, for which application of end-of-the-pipe treatment technologies alone will be impractical and costly.

Similarly, some environmental problems involve firms that are heterogeneous with respect to geographic dispersion, age, control options, and competitive position. Generally, these kinds of problems are not amenable to resolution using traditional command-and-control approaches. An example is the persistent problem of non-point source water run-off. Since run-off comes from both urban and agricultural sources, the cost of control varies substantially. Because the pollutants are both diverse and have multiple impacts, design standards and individual equipment performance standards fail to find cost-effective and administratively homogeneous regulatory solutions.

It is clear there is no one-size-fits-all regulatory solution. Command-and-control is the right regulatory approach for certain problems, whereas one of several economic instruments might be preferable for others. While the problem of non-point source water run-off is unlikely to be solved cost-effectively through a command-and-control approach, a standardized, one-size-fits-all economic incentive approach will not work either. In all cases, the regulatory solution must be custom-fit to the problem.

One goal of pollution control is improved health, for example, minimization of the unhealthy effects of pollution. An immediate question economists ask and regulators must deal with is, what costs should we impose on society to improve health? Another question is, how do we handle the different impacts of the costs on differing firms? Both command-and-control and economic incentive-based regulations must deal with these problems. Economic incentive-based regulations, however, have the great advantage of being easier for regulators to implement because regulators need not make guesstimates as to what is the right technological fix. Instead, economic incentive approaches leave that decision to affected firms and the marketplace. Given the wide diversity in both the costs of compliance and resulting environmental benefits, it has often proved difficult to design simple and enforceable design-based regulations that meet this standard.

Experience has shown that the most effective time to use direct regulatory, or command-and-control, approaches is when the pollution emitted by numerous sources is homogeneous and regulators can identify control technology that achieves the desired standards at a relatively low cost.

Along with cost-effective design concerns, legitimate monitoring and enforcement concerns must be considered if a program is to result in the kinds of outcomes regulators initially project. Cost-effectiveness, monitoring, enforcement, and equity concerns must be married to get the best outcomes. Doing so, however, means that regulatory programs will, by necessity, be composed of elements of command-and-control, economic incentive, voluntarism, and other regulatory models.

Given that industry is not homogeneous in terms of production processes, and given that the nature of environmental regulation has been changing, many regulators and regulatory analysts are encouraging a shift away from traditional regulatory models. Proponents of this shift argue that regulators need to be able to tailor the ambient standards and deadlines to the differences in regional conditions. Regardless of the merits of that argument, critics of the status quo contend that more-goal-oriented and market-based approaches will help achieve statutory goals in a timelier, less costly, simpler, and more flexible way.

Finally, the traditional regulatory approach has been costly. All levels of government have spent substantial resources to develop, administer, and enforce design-based regulatory programs. In addition, the private sector has spent great amounts of money to comply with the regulations. The cost of future environmental improvements will likewise be high. Study after study concludes that by using the power of the market and economic incentive-based regulation, the private sector can obtain more pollution control at less cost.

THE ADVANTAGES OF ECONOMIC INCENTIVE-BASED APPROACHES TO ENVIRONMENTAL CONTROL

The limitations of the command-and-control regulatory approach, complexity of remaining and emerging problems, pressure on the industrial sector to keep prices low to be internationally competitive, and the persistence of budget deficits have led many regulatory analysts to call for greater use of market-based programs for environmental control. Many observers argue that for the United States to progress toward its environmental goals, it must move beyond the traditional technologically based prescriptive approach and add innovative policy instruments to the menu of options available for environmental management. The following case studies and EPA's acid deposition control program provide powerful illustrations of the benefits of these innovative regulatory approaches. In using these approaches, the first issue that has to be addressed is how to select which approach is appropriate in a given area. The second is how to implement a certain mandated program successfully, given the minimal attention typically paid to implementation.

Different localities are now using economic incentives to address the above types of problems. Economic, or market-based, programs rely on pollution charges or marketable permits as leverage to motivate pollution sources to seek ways to limit their pollution that are consistent with their self-interests while meeting environmental goals. Properly employed, economic incentives can be a powerful force for the mitigation of certain environmental problems.

To some, using incentives to protect the environment is a radical concept compared with the traditional regulatory methods of protecting public health and public welfare through direct dictates by a presumably omniscient regulator. However, almost 20 years of pilot programs suggest that economic incentives have worked well as both a complement to and substitute for command-and-control-based environmental programs. Particularly in situations where the environmental effects are not acute, greater consideration of cost-effectiveness may be warranted. In addition, many environmental problems may be too intertwined with everyday activities to be solved through centralized regulatory systems.[1] To control hydrocarbon emissions from coatings and deodorant may be difficult if reliance is placed on a point-of-use control technology. If, however, the hydrocarbons in the product are taxed, industry will seek company-specific solutions.

The rationale for market-based approaches is straightforward. Direct or indirect price or cost signals aimed at pollution are used to encourage individual decision-makers to determine what action is best, given their unique circumstances. Some sources may prefer to pay the governmentally imposed market costs and continue to pollute, or they may choose to install equipment instead of buying marketable permits. Others may find it cheaper to modify their production processes or input products in ways that eliminate or reduce the pollution, thus avoiding or minimizing the external costs associated with the pollution. The totality of the individual responses to the incentives should result in a reduction in pollution at a cost that is lower than the alternative, since only those polluters that find it cheaper to do so will undertake abatement. The higher the price for polluting (ceteris paribus), the greater the incentive to reduce pollution is, and therefore the greater the reduction is likely to be. The cost-effective properties of economic incentive systems have been widely

---

[1] Those systems will never have as much data on product or process substitutes as the private sector. In this context, regulators must find ways to leverage the knowledge of the private sector and create incentives to meet social goals.

accepted in theory and practice. Today the principal questions are how to design and administer such systems.

In summary, incentive- or market-based options offer a number of advantages:

(1) **Carrots and sticks.** As noted, market-based approaches give polluters a financial reason to reduce pollution, usually with maximum flexibility as to how to achieve those reductions. Rather than direct enterprises on how they should control their pollution, incentive-based systems impose a cost on pollution-causing activities and leave it to individual firms to decide for themselves (jointly or individually) how to achieve the required level of environmental protection. With an incentive mechanism in place, the polluter pays a financial penalty for high levels of pollution and pays a lesser penalty or receives a financial reward for lower levels. In theory, incentive mechanisms can reduce pollution at less cost than traditional regulatory means.

While subsidies are included in incentive-based approaches, and can produce many of the desired results achieved by taxes and marketable permits, they are not completely consistent with the polluter-pays notion. Still, subsidies, when used with other tools, may many times be superior to a simplistic command-and-control approach.

Performance standards also relieve the regulator of the responsibility to dictate specific control strategies. Performance standards capture many of the economic efficiencies that economic instruments do. In fact, combining performance standards with marketable permits or pollution charges may be a powerful control strategy.

(2) **Cost-effectiveness.** The most economically efficient incentive is theoretically the one that requires the polluter to pay the exact price for the pollution, valued on the basis of its damage to others. Paying more than the societal burden is not cost-effective. Also in theory, the source should reduce its pollution to the point that the cost of further reductions exactly equals the incremental damage its pollution causes others.

Incentive mechanisms have several properties related to economic efficiency that can make them especially well suited to some of the environmental problems the nation faces now and in the future. First, relative to traditional forms of direct regulation, incentive approaches allow regulators to deal more effectively with pollution from diverse sources—an increasingly important problem. Second, for many types of environmental problems, incentive mechanisms may be more economically efficient; that is, they achieve environmental goals at lower cost than does direct regulation that specifies the technologies to be used.[2] Third, incentive mechanisms provide a greater stimulus for cost-effective innovation and technical change in pollution control than direct regulatory approaches do.

---

[2] See Robert Hahn and Carol May, "The Behavior of the Allowance Market Theory and Evidence" *The Electricity Journal*, March 1992, Vol. 7, No. 2. and C. Jubb and B. Underhill, "Valuing the Environment: Theory, Methods and Proposed Application," BIE Working Paper 59 (1990).

(3) **Clarification.** Incentive-based approaches can make the environmental debate clearer to the general public. Economic incentive-based regulatory programs take as a given the ambient or effluent reduction goal. Thus, they allow attention to focus directly on the issue of what the environmental goals should be. In contrast, regulations dictating a particular technology draw attention to difficult questions regarding the technological alternatives for reaching those goals.

(4) **Socially desirable outcomes.** Because incentive policies establish incentives for sources that reflect important social as well as private costs, they encourage private decisions that more closely approximate socially desired outcomes as construed by the designers of the incentive-based system. For example, under Title IV of the 1990 Clean Air Act (CAA), Congress made the decision that there should be a 10 million-ton-per-year roll-back of sulfur oxides (SO_x) coming from powerplants. Congress frequently makes such choices. Once the choice has been made, however, the means of achieving this goal is put into the hands of the private sector. Under an incentive-based system, the public sector focuses on the goal, while the private sector focuses on meeting the goal. Under command-and-control-based systems, the public sector is involved in both the end and the means. While government might get the goal-setting right, it rarely determines what the best choices are for the individual units it regulates. As such, only rarely do prescriptive command-and-control regimes achieve both desired environmental and economic outcomes.

(5) **Administrative simplicity.** Incentive systems are designed to achieve low cost and administratively simple outcomes. In contrast, command-and-control schemes require detailed knowledge of many industrial processes or other polluting activities. Regulators need this knowledge so that they can develop regulations that call for specific pollution control technologies that can achieve needed reductions in pollution. Getting the data required to develop effective design standards is a long and expensive job. Moreover, the process is inherently uncertain because the private sector is unenthusiastic about collecting data on control costs before regulations are developed. Industry is even more unenthusiastic about giving these data to regulators.

A second reason incentive-based systems are administratively simple is that they require much of the regulatory energy to be expended up front, in the design stage of the regulatory program. If the design is correct, less burdensome administration may be facilitated. Further, once the program is in place, regulators can rely on the energies of the private sector to drive pollution downward. Command-and-control regulation, on the other hand, may impose a never-ending requirement on regulators to develop new and more stringent industry-specific regulations on smaller and smaller discharge points. That administrative burden is one reason regulators in the Los Angeles area adopted a marketable permit system for achieving reductions in nitrogen oxide and sulfur oxide (see the section on RECLAIM [Regional Clean Air Incentives Market] in chapter 2). Simply put, Los Angeles air regulators did not have the resources or time to develop source-specific regulations continually. The marketable permit system allowed them to shift the responsibility for identifying optimal control strategies to the private sector.

Despite the theoretical appeal of incentive mechanisms, they are adopted with a certain degree of trepidation. One reason is the level of comfort and legal experience that some federal, state, and local regulators have with command-and-control regulatory systems. Another reason is that, as experience has shown, the regulated community may roll back some environmental goals in the process of adopting market-based programs. In other words, some ground may be lost initially. The change involved in entering into untried programs produces fear. The resistance can be particularly strong where the change challenges the traditional view of the organization's mission and how it achieves its goals. At the core of traditional regulation is a belief in the dedication, competence, and beneficence of the civil servant. Economic incentive-based systems challenge this belief—they are based on the assumption that the private sector knows more about control opportunities, can work faster, and is more open to finding the best system than even the most knowledgeable, benevolent, and dedicated regulator.

There is also an erroneous perception that efficiency and the attainment of environmental goals are incompatible. While in many instances they may conflict, by selecting the right regulatory tool for the problem, regulatory architects can minimize conflict.

Next, the command-and-control culture is well-established in many regulatory agencies, a condition that contributes to the substantial mythology surrounding some of the enforcement and administrative benefits of these programs. Although never substantiated, that mythology makes some regulators, already antagonistic to reform, believe that the traditional is superior in many ways. In short, in a number of public organizations the status quo is a vicious competitor against any new idea.[3]

Finally, two other obstacles hinder the use of market forces and economic common sense to achieve environmental goals. One is the barriers to market forces imposed by legal limitations. The second is the barriers imposed by other government programs, such as the tax code, subsidies provided by government agencies, or other provisions of existing laws or regulations that create imperfections in the marketplace. For example, the tax code can significantly influence how firms trade marketable permits; subsidies for specific pest management practices or crops can influence the pesticide and fertilizer use by growers; and subsidies for so-called clean coal technologies can distort the results expected from a market-based regulatory program predicated on simple models of rational decision-making by regulated entities. To the extent that a goal is to solve problems in an economically efficient manner, these obstacles need to be overcome.

---

[3] Although proposals vary, most proponents of incentive approaches see them as complementing existing environmental regulatory structures rather than replacing them. However, incentive-based regulatory programs can form the foundation for a regulatory regime, consistent with Title IV of the 1990 CAA Amendments.

Removing unwarranted subsidies and hurdles, or a failure to specify particular technological solutions, does not mean, however, that the federal government should abandon the provision of all goods and services to the marketplace. The federal government is not just a regulator. It is also an important owner of public lands and is the custodian of bodies of water. It has a wide array of responsibilities outside the field of regulation.

Finally, there is the issue of theory versus practice. In theory, incentive mechanisms will reduce the cost of pollution control because sources will try to minimize the total cost of pollution control. Under an optimally structured incentive system, profit-maximizing firms control pollution until the marginal costs of control are equal for each source. In practice, however, the design of an incentive system may be sub-optimal. For example, to be politically viable, an incentive-based system may have to exempt some sources of pollution. Exemptions may be the price that a regulator must pay to gain political support for the regulatory programs. Further, the program may calculate the fees or permit totals that are the foundation of the program incorrectly, it may not use efficient criteria, or some underpinning of the program might be incorrect. For example, firms may minimize risk instead of maximizing profit. In that case, the outcome of a program predicated on profit-maximizing by firms will be not be obtained. In these cases, an incentive system will approach but not necessarily achieve maximum possible savings.

Assuming all these issues can be addressed, an economic incentive-based system should be able, simultaneously, to:

- Foster environmental protection.
- Promote a stronger, more competitive economy.
- Promote cost-effective innovation in pollution control technologies.
- Promote prevention of pollution.

THE NATURE AND APPLICATION OF ECONOMIC INCENTIVE PROGRAMS

As noted, environmental incentive systems are mechanisms whereby market forces establish a price for or limit the quantity of pollution.[4] They influence rather than dictate the actions of the targeted parties. They leave the ultimate choice of what action to take to the affected parties, based on their own evaluations of costs and benefits. Different incentive mechanisms use different methods to establish the prices for pollution. However, in all cases, those who

---

[4] This definition excludes some mechanisms that might be considered incentives. For example, fines discourage littering. Under the conventional definition of an economic incentive, a fine may not be considered an incentive mechanism. The reason is that flat littering fines do not vary with the amount of litter. A polluter pays no positive price for each unit of additional litter. A price exists only in an expected value sense over repeated events. Nevertheless, penalties are part of the tool kit for economic incentive programs.

pollute pay a price for each unit of pollution and those who reduce pollution receive benefits for the number of units of pollution they eliminate.

Typically, incentive systems are applied to intentional permitted discharges. They can also be applied to accidental or unintentional environmental effects or pollution. For example, under the Comprehensive Environmental Response, Compensation and Liability Act (CERCLA), the person or party responsible for an accidental release of pollution is liable for the total costs of responding to the damage and of the damage itself.[5] Some incentive mechanisms affect prices directly, examples being charges based on the volume and toxicity of effluent discharges, pay-per-bag systems for solid waste disposal, permit fees for air emissions (in which the fee varies with the type of pollution or volume emitted), and refundable deposits on containers. Tinkering with prices can bring about many changes in firms' polluting behavior.

While some advocates of reform promote economic incentive-based regulatory programs to increase the cost of pollution and thereby change behavior, some regulators have sought to use pollution charges as a revenue-enhancing tool to cover the costs of processing permits and other activities associated with regulation. The Russian pollution charge system, for example, generates funds that can be employed for a wide variety of environmental needs. Raising revenues for fiscal purposes is, however, usually not considered a rationale for economic incentive-based regulatory programs.

Experience with incentive systems raises some issues that need to be addressed in implementation:

- *Incentive systems may require additional monitoring and enforcement as specified by regulators, thus increasing the administrative burden.* Monitoring and enforcement have been among the thorniest issues policy-makers have had to deal with in connection with economic incentives programs. While sometimes the monitoring and enforcement requirements under the two approaches differ, there is no evidence that monitoring and enforcement should be more difficult under incentive programs, particularly with today's technology. Both systems are predicated on permit systems that aim to capture reliable data and have enforceable discharge limits. Without a good system of permits or good monitoring systems, both systems may fail. To the extent that there is a difference, however, more alternatives to assure compliance may be available under an incentive-based system, especially if the compliance assurance technique is based on performance measures and not on design criteria.

- *The appeal and acceptability of an incentive system in comparison with*

---

[5] Liability systems, which are also part of tort law, are excluded from the present discussion, largely because they involve identifiable harm and identifiable polluters. In contrast, the incentive systems of interest here are by and large applied to situations, such as automobile exhaust pollution, where it is difficult or costly to determine the exact connection between an individual polluter and the impact on the environment.

a command-and-control system depend greatly on who gains and who loses from its use. Those who gain are likely to favor incentives and those who are made worse off are likely to oppose them. This issue is complicated by the fact that overall gains across firms may be substantial and individual gains small, and losses may be concentrated among just a few firms. Unfortunately, the power of economic incentives is in the cost-effectiveness provided to many firms. This means some potential inequities from the perspective of potential "losers." Thus the debate within industry takes an efficiency-versus -equity orientation.

- *There has been a tendency (from industry's perspective) on the part of some regulators to use regulatory reform as an opportunity to micromanage compliance by industry.* As a result, some firms look at some economic incentive-based systems as imposing more administrative and monitoring costs on both industry and regulators than the system being replaced. That outcome imposes hidden costs on both industry and regulators and can result in lower cost savings and frustration on the part of stakeholders. For example, under the acid rain control program, utilities are required to install continuous emission monitors. This monitoring process results in an increase in cost and in a greater administrative burden than do current command-and-control regulations. As stated, some industries wait to use the movement to market-based systems as a means of avoiding cleanups.

- *When calculating the distributional impacts of an incentive system, it is necessary to begin with the mechanism itself.* Compared with command-and-control, a party may feel worse off when a tax is imposed, other things being equal. Similarly, parties that have to buy marketable permits may feel poorer compared with how they would feel under a command-and-control system. While these "feelings" must be recognized by regulatory designers, and these feelings may influence the final design of a system, regulatory designers must be able to deal substantively and rhetorically with perceptions by individual firm managers and companies when explaining the costs and benefits of incentive-based systems. The RECLAIM case study and comments from industry illustrate how firms can perceive that they are losers while industry as a whole is a winner.

SELECTED INCENTIVE-BASED ENVIRONMENTAL CONTROL POLICIES

The idea of charging industry a price for disposing of waste products into the air and water comes from the welfare economics teachings at Cambridge University in England early in this century. Beginning in the fifties and sixties, economists refined and elaborated this basic concept into various proposals for pollution charges, a mainstay of conventional economic thought on how to manage the environment. A number of pollution charges have emerged, including discharge fees, whose level is based on environmental damage; direct taxes on discharges; and tradable permits and emissions trading. A fourth charge-type program that emerged in the last 20 years is the beverage container deposit-return system.

Discharge Fees

Under this approach a regulatory authority imposes an emission or effluent fee on the discharges from individual pollution sources in proportion to an estimate of the environmental damage caused by the discharges. The rationale for discharge fees rests on two fundamental arguments. First, price signals promote the efficient use of scarce abatement resources. Second, discharge fees act to ration the limited assimilative capacity of water bodies and airsheds. While in the past some analysts thought the solution to pollution was dilution, understanding how much pollution can be assimilated by the environment during a given time period is the first step toward either a command-and-control approach or an economic incentive approach to pollution control. The rationale for action at the regional level is that, to determine damage, the geographic area covered must be limited, as damage varies according to such factors as climate, ecology, and population density. In this way, by tracking the ambient impact back to the source, the social costs of the polluting behavior become part of the source's cost of doing business. In the parlance of economists, the external costs of pollution are internalized.

Direct Discharge Taxes

The approach of direct taxation of pollution has two clear advantages: it ensures a direct source of revenue for general government or for earmarked activities; and it provides a disincentive to engage in polluting activities. It may also help government avoid having to raise traditional taxes such as income, property, or sales taxes. The case study on the Russian pollution charge system is a perfect illustration of using a tax for both regulatory and financial reasons.

Economists frown on using a single instrument for too many purposes and are concerned that the efficiency and equity of a single targeted instrument get lost when trying to achieve multiple goals. While taxes have both efficiency and equity characteristics, the issue is the difficulty of using one tool to meet multiple objectives. Perhaps it is better to use a variety of tools to meet a variety of goals.

Taxes can be leveled on rates or total volumes of discharges. Properly designed, the tax offers a direct incentive to sources to find ways to reduce or eliminate the cost.[6] Sufficiently high taxes encourage quick action and so may induce sources to work faster than they would under traditional regulatory

---

[6] In contrast, under a discharge fee system, the charge is independent of location. As such, it could be less complex to implement.

approaches. A sufficiently high tax is one that is greater than the incremental abatement costs some polluters face. Setting the level of the tax is likely to be an iterative process, since actual reductions are hard to predict in advance. Therefore, many steps might be required before the right tax level is found. Note that the revenues would not be used for subsidies, since doing so would defeat the general revenue objective. The Russian pollution charge system illustrates the use of a tax system that has one tax rate for discharges at one level, a second rate for discharges at a second and more threatening level, and a third, very stiff rate for discharges at a still higher level.

Tradable Permits and Emissions Trading

Tradable permits for discharges into the water or air are marketable commodities that allow polluters to discharge specified amounts of pollution into a water system or airshed. The total amount of discharge is based on regulatory targets and the disposal capacity of the water body or airshed. If the authorities set the total amount of pollution allowed under the sum of the permits at less than existing pollution levels, the permits are scarce and may be quite valuable elements in a firm's portfolio of productive inputs. These permits are allowed to be traded among firms. Firms whose cleanup costs are less than the market price of the permits may decide to sell their permits, while firms whose costs are greater than the price of the permits may decide to buy them. In theory, the combined effect is a least-cost reduction in pollution that meets ambient quality goals without specifying in detail how firms are to do so.

The focus of tradable permit schemes is on limiting the quantity of discharges, rather than on price directly. In theory, pollution charge systems and marketable permit-based systems will achieve the same environmental and economic outcomes. In practice, however, they may not. Relative to fee- or charge-based systems, trading systems may cost polluters less, since, under existing practice for allocating pollution rights, the maximum cost to polluters is the cost of meeting the regulations or standards set to control the pollution. With fee- or charge-based systems, on the other hand, initially environmental goals may provide the drive to change behavior but later may be used for such purposes as supporting regulatory agencies and promoting technology. As a revenue measure, the charge could become greater than needed to achieve the optimal levels of discharges.

Trading systems have a number of important attributes, including a well-defined scope of coverage and a clear focus on the technical basis and geographic limits for the trading. The programs can involve inter-firm trade and intra-firm trades between product lines and between locations. The programs can be local, national, and international in scope.

A trading program can involve either credits or allowances. A credit is created by a source that emits less than its allowable limit. To obtain the credit, a polluter is required to demonstrate that the reduction is surplus and meets other regulatory tests. Regulators grant credits when reductions are below the regulatory baseline. In a credit program, a designated authority must certify the creation of the credit and the trades. It is clear that a credit-based system gives regulators two opportunities to regulate the creators of credits. The first is when the baseline and ground rules are established, and the second is when firms apply for emission credits.

In a sulfur dioxide (SO_2) allowance trading system under Title IV of the CAA, the regulator grants quasi-rights or allowances to emit, and the trading involves future pollution. Firms are "allowed" so many tons per year to pollute; if a firm does not need all of these "rights," it may sell them. Once EPA sets an allowable limit for a source, the source can add to its allowable limit by acquiring allowances. If a source controls more emissions than needed, it generates excess allowances that it can save for future use or can sell. The agency should, at a minimum, record the trades, but it need not certify each transaction involving an allowance. The certification of allowances for each source takes place before trading and may be revised whenever a source changes its pollution control equipment. The description preceding this sentence also applies to the NO_x trading program under RECLAIM.

Tradable discharge permits offer cost-effective incentives similar to a discharge fee system without some of the administrative problems. Depending on how tradable permits are initially allocated to polluters, they can also overcome some of the financial impact on the sources of the polluter fees. Emissions trading has also allowed greater flexibility in the current regional air quality programs of some states. Because most environmental control regulations already involve permit systems, this approach appears to be a natural supplement to existing institutions.

Although all trading programs require some involvement by a pollution control agency, the nature and level of that involvement vary substantially across programs. All programs require that trades be recorded (although not necessarily by government), and in some programs, recording the trades may be the agency's only role. In other programs, the agency has to approve every trade. Whether the agency has to approve every trade is a major variable across programs. Agency approval is likelier where the trades involve different pollutants that have different locational impacts than where the pollutants are measured differently or where the permit systems are weak or embryonic. Certainly any trading program involving toxics will require a great deal of regulatory oversight, whereas trading of nuisance pollutants requires less. Spatial and toxicity issues are of concern to regulators, environmentalists and industrialists in the case of both air and water programs. Regulations that clearly spell out the terms and conditions for dealing with spatial and toxic concerns are used to address these issues, as is evident from the U.S. and Russian case studies. How these problems will be dealt with in emerging programs will be of great interest.[7]

---

[7] For example, how trading or taxes might be used to eliminate persistent toxic substances in the Great Lakes is being reviewed, with no answer as yet.

Deposit-Refund Programs

Deposit-refund programs address two costs that usually are external to beverage manufacturers, distributors, and consumers—disposal and littering. Deposits provide a disincentive to create specified types of litter and an incentive to collect litter and reduce the volume of solid waste. That is, the goal of a beverage container deposit-refund system is to encourage the collection of beverage and perhaps other containers for recycling, so as to reduce litter and potentially reduce the quantities of solid waste disposed of, as well as to reduce the energy use and pollution associated with the extraction and use of virgin materials. Requiring a small refundable deposit on a product or substance provides the incentive for recycling or return, so that the volume of waste is decreased. Bottle bills are largely self-implementing. Once the deposit is collected, there is an incentive on the part of consumers and retailers to return the containers to reclaim the deposit.

As expected, mandatory deposit mechanisms are somewhat limited in their application. They can be used only for products and substances that are either durable or reusable, are not consumed, or are not dissipated during use. The most common use for deposit-refund systems is mandatory collection of bottles. It should be noted that mandatory deposit-refund systems are also sometimes used to protect the environment against hazardous substances (for example, pesticide containers, auto batteries, and small quantities of hazardous wastes) that cannot be recycled. In these cases, deposit systems attempt to achieve cost-effective disposal. In the case of hazardous waste, the system seeks to prevent what is commonly referred to as midnight dumping by creating financial incentives for the proper recycling or disposal of wastes.

Although national legislation mandating the use of deposit-refund systems does not exist, most states using such approaches have had high rates of success with returns. Those with mandatory deposit legislation argue that it has reduced litter significantly. In general, mandatory deposit-refund systems reduce the burdens associated with monitoring and enforcement and return valuable materials to manufacturers for recycling.

SUMMARY

Command-and-control and economic incentive-based modes of regulation can be both supplements and complements to one another. Each has a proper place in the regulator's policy tool kit. The question before both legislators and regulators is always how to get the best outcomes, at the lowest cost, with the least administrative inconvenience to both the private and public sector, with the greatest influence on long-term innovation in pollution control technologies, and in the most enforceable way. Neither system is a panacea. Both have a role. The following case studies illustrate the evolution of four different incentive-based systems and point out some of the design and administrative problems and successes associated with this approach to regulation.

2
MARKETABLE PERMITS AND POLLUTION CHARGES: TWO CASE STUDIES

When economists think about economic instruments, usually they think of activities that influence either the price of a product or service, or they think of an activity that influences the available quantity of that product or service. The reason is that the fundamental tool of economic analysis is the assessment of trade-offs between price and quantity. In general, as the price goes up, consumers demand less of a given product. Still, producers are eager to provide more. Contrariwise, as the price of a product declines, consumers are eager to buy more. However, suppliers are uninterested in producing more. This inverse relationship between the interests of consumers and suppliers is a fundamental focus for economic analysis. (Box 2-1 presents a description of these trade-offs.)

Pollution charges are like taxes in that they add to the cost of a product. Influencing the price of the product in turn influences the quantity of the product. As noted, as the price of the good increases, the relative attractiveness of substitutes also increases. The substitutes will get a great share of the market unless producers lower the profit margins for the taxed product, which some producers may be willing to do to secure market share or achieve other goals.

Similarly, if government limits the amount of a good available to the market by restricting its supply, the price of the good may increase. There are many examples illustrating how governments limit the supply of certain products or services. In many cities, there are limits on the number of taxicab medallions. If the taxicab business becomes increasingly profitable, the price of the medallions will rise. If the business falters because of the availability of cheap public transportation, the price will decline.

Economists have encouraged the use of taxes, sometimes called pollution charges, as a way to limit the supply of pollution. If the tax is imposed on, for example, carbon emissions, producers will be inclined to lessen carbon emission. In fact, they should be inclined to install either pollution control devices or other control measures up to the point at which the tax is equal to their cost of controlling the emissions. By raising the tax, regulators can increase the cost of polluting and make pollution control activities more attractive.

> **BOX 2-1**
> **ECONOMIC THEORY UNDERLYING MARKET INCENTIVES**
>
> Economists like to work with supply and demand curves. A supply curve is a schedule of quantities supplied by producers under different pricing regimes, and a demand curve is a schedule of purchasing decisions made by consumers under varying pricing regimes. The point at which the supply and demand curves intersect identifies the amount of a product produced.
>
> A tax or pollution charge moves the demand curve because it adds to the cost of products (through a direct tax or through the cost of buying marketable permits). At any price, under the new regime less of a product is purchased. Therefore, if the government adds a pollution charge to the cost of producing pollution, businesses will produce less of the pollutant because they will find it hard to pass the cost on to purchasers of their final product, for example, steel, radios, or chemicals.
>
> Economists argue that since it is difficult for even the most knowledgeable regulator to know exactly the right technology for every polluter to install, if the government raises the price of pollution through quotas or discharge fees, industry will seek the most cost-effective solution for each and every facility. By focusing on prices and quantities, it is possible to focus on a fulcrum against which to leverage many firm-level decisions and promote demand-oriented innovation, instead of regulatory-driven supply-oriented technology innovations.
>
> Understanding the shape of supply and demand curves for pollution is an important aspect of designing regulatory systems. In addition, understanding the motivations of industry, the ability of industries to pass costs on to consumers, and industries' profit-maximizing, cost-minimizing or risk-minimizing attitudes toward pollution control can give attentive regulators important insights into the design and implementation of economic instruments.

Likewise, regulators could limit the annual tonnage of carbon emissions, distribute permits (as they do taxicab medallions and liquor licenses), and then let the marketplace establish a price for the reductions. Firms that find it cheap to control carbon emissions might over-control their emissions and sell "extra" reductions to others for which carbon controls are expensive. By sequentially lowering the limits on carbon emissions, less and less emission is allowed and all else being equal, the cost per permit should increase.

In both cases—either by tinkering with prices or by limiting the quantity of emission permitted—regulators adjust the "market for emissions" to reach the socially acceptable ambient target determined by separate scientific analysis.

There are many reasons a regulator might choose to use quantity-limiting quotas to achieve a regulatory goal. They include:

- Quantity goals may have to be reached within a certain period.
- Quantity relationships are better understood than pricing relationships.

- The legal authority to tax is absent.
- The history of the regulatory program dictates quantity-limiting measures.
- They are compatible with other existing and forthcoming regulatory programs.

Pollution charges to support regulatory objectives are preferable to quantity-limiting quotas when:

- The quantity relationships may be uncertain, and taxing appears to be more acceptable politically.
- Property rights are not well-defined, so that using quantity-limiting tools is ambiguous within the legal framework.
- Revenues can be used to offset the regulatory costs.
- Revenues can be recycled to achieve other social goals.

Russia and much of Eastern and Western Europe have employed pollution charges to influence the cost of polluting and produce the desired environmental outcomes.

Because of a reluctance in the United States even to utter the "t" word—taxes—and because of how U.S. air and water pollution control laws have been designed, regulators have developed marketable permit-like systems. In addition, it is difficult to establish appropriate charge levels because little is known about the health and welfare costs of given pollutants. The first air credit trading program, called the offset policy, was developed in 1976. This policy became the basis for all the air credit trading policies and programs that followed—the bubble policy, emissions banking, netting, lead trading, sulfur dioxide (SO_2) allowance trading, and the Regional Clean Air Incentives Market (RECLAIM) program. While in the United States pollution charges are used in conjunction with a variety of permit fee programs, currently the fees are not large enough to alter behavior in ways that can be measured.

The two case studies that follow—the marketable permits program of the South Coast Air Quality Management District (SCAQMD) and the Russian national pollution charge system—demonstrate different paths to achieving similar goals, such as lowering pollution and creating a healthy environment. The Russians are using discharge fees, the Americans marketable permits; the Russians established their system after a series of experiments, while in the United States almost all the effort in developing each policy went into design, with little going into field experiments; the Russian program encompasses many pollutants, the U.S. program a few air pollutants; and the Russian program is still embryonic, while much of the U.S. program is mature. Despite these differences, the two case studies offer some similar lessons—regulators paid little attention to establishing good oversight systems to provide for midcourse corrections; the importance of good monitoring and enforcement cannot be overstated; and a tenacious champion of regulatory reform can move an idea forward even in the face of opposition or indifference from stakeholder groups.

CASE STUDY:
SOUTH COAST AIR QUALITY MANAGEMENT DISTRICT AND THE MARKETABLE PERMITS PROGRAM

The SCAQMD is responsible for bringing the South Coast Air Basin into compliance with the air quality standards established by the U.S. Congress, EPA, and California Air Resources Board (ARB) (see box 2-2 for the origins and boundaries of the SCAQMD and figure 2-1 for a map of the boundaries). The District has a wide array of powers and responsibilities and has been using the entire range of policy tools in implementing federal, state and local regulations. These tools include direct command-and-control regulations, permit fees, marketable permit programs and other incentive-based systems. The SCAQMD's responsibilities include: developing Air Quality Management Plans (AQMPs); establishing regulations and rules to implement the plans; applying the rules; and prosecuting non-compliance through fines and civil or criminal proceedings. The District has a great deal of autonomy in establishing regulations and rules and in setting and imposing fees and fines.

The SCAQMD is notable for its use of economic incentives as a foremost means of promoting compliance with air quality standards. It has emerged as

---

BOX 2-2
THE SOUTH COAST AIR QUALITY MANAGEMENT DISTRICT

In response to federal requirements for improvements in air quality as specified in the Clean Air Act of 1970 (CAA), on February 1, 1977, the California state Legislature created the SCAQMD, which combined the Air Pollution Control Districts located in the South Coast Air Basin. The South Coast Air Basin covers a large area in southern California, bounded by the Pacific Ocean on the west and the San Gabriel, San Bernardino, and San Jacinto mountains to the north and east. The Basin encompasses topography and climate —light winds, a heat inversion layer, and substantial sunlight— that together create an area of high pollution potential. These natural conditions support a large population (measured at 12 million in 1987) that is growing at roughly 1.2 percent a year, a diversified industrial economy, and the largest urban concentration of automobiles in the world. Transportation accounts for more than half the emissions of five important pollutants. The industrial and commercial sectors together cause 42 percent of the volatile organic compound (VOC) emissions and declining shares of the others. The residential sector generates 11 percent of the VOC and insignificant portions of sulfur oxides (SO_x) and particulates (PM10). Within the Basin, the SCAQMD covers the city of Los Angeles, Orange and Riverside Counties and the non-desert portion of San Bernardino County.

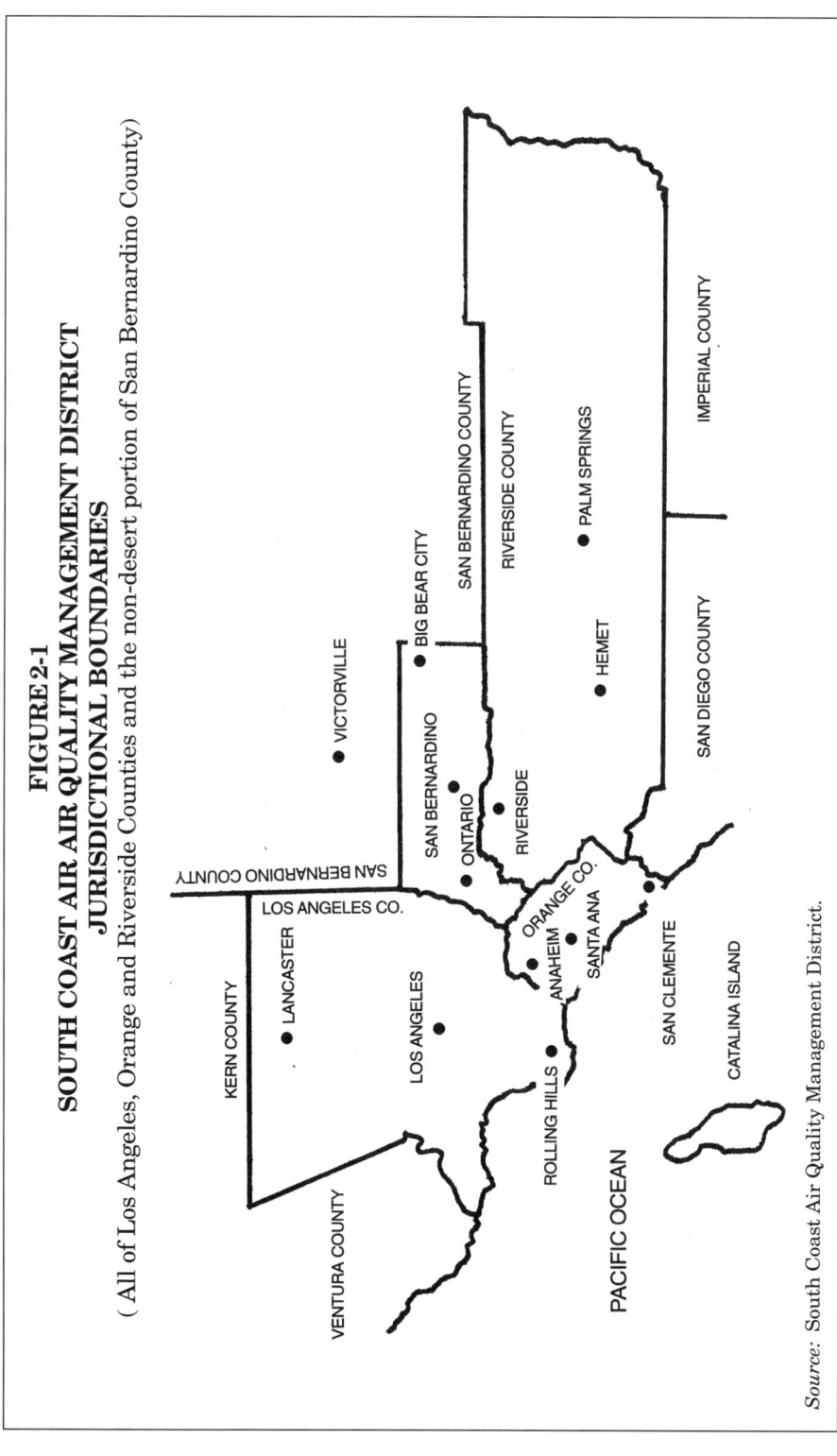

**FIGURE 2-1
SOUTH COAST AIR AIR QUALITY MANAGEMENT DISTRICT
JURISDICTIONAL BOUNDARIES**

(All of Los Angeles, Orange and Riverside Counties and the non-desert portion of San Bernardino County)

Source: South Coast Air Quality Management District.

the most active emissions trading market in the country. It has successfully implemented an air credit trading system. One outcome has been the development of a knowledgeable business and environmental community familiar with many air credit trades. These groups have been able to work effectively together, despite considerable strife, in the pursuit of regional air quality goals through the use of economic incentives, first with the offset policy and presently with a new market-based program called RECLAIM.

The SCAQMD is now taking on an unprecedented challenge as it seeks to shift from primary reliance on command-and-control regulation to a wholesale reliance on market-based regulation to meet nitrogen oxide (NO_x) and SO_x attainment goals. The focus of much of the regulatory operations in 1992-93 has been the development of RECLAIM, an alternative regulatory system aimed at attaining the ambient standards associated with NO_x and SO_x from stationary sources. If successful, RECLAIM will result in reductions in emissions by approximately the amounts required by regulators—75 percent of NO_x and 60 percent of SO_x. It will do so by mandating emission reductions and then leaving it to industry to decide how to accomplish the reductions. The SCAQMD is considering the development of a manufacturers' bubble program, another market-oriented initiative. RECLAIM is a significant departure from the traditional strategy of the SCAQMD. The program is designed to provide maximum flexibility to sources while stimulating innovation and advances in technology.

This chapter looks at the evolution of economic incentives in the SCAQMD and at the SCAQMD's ongoing evolution from a regulatory agency that has relied almost exclusively on command-and-control approaches to one that is about to implement a novel market-oriented approach to environmental management. This shift has necessitated profound changes in the thinking and operations of the SCAQMD, the companies being regulated, and the public and national regulatory agencies. To understand fully the difficulties associated with the move toward RECLAIM, this case study begins with background on the mandates and processes established by the CAA and emissions trading under Regulation XIII, the economic incentive program that established the regulatory framework for RECLAIM. Second, it describes the organizational structure of the District and the District's budget. Third, the South Coast banking and trading market is reviewed, including the industry's perspective. Finally, there is a detailed description of the development of RECLAIM, including the structural, administrative, and operational changes required at the SCAQMD; the costs of the program; and the potential challenges associated with its implementation.[1]

---

[1] To obtain the information on RECLAIM, a team from the National Academy of Public Administration (NAPA) interviewed individuals involved in its development, both inside and outside the District, during the winter and spring of 1993 and reviewed published reports, budget documents, meeting notes, and other written materials provided by District staff and other parties.

THE U.S. CLEAN AIR ACT

The primary objective of the CAA is to protect citizens and natural resources from pollutants that could endanger public health and welfare or destroy the nation's natural resources. The CAA dictates how ambient standards are determined; the distinction between clean-air and dirty-air regions; which emissions are to be regulated; how states are to regulate air pollution; procedures for writing and issuing permits for sources that emit air pollutants (air permitting procedures); how sources of air pollution are defined and monitored; and how compliance is assessed. To this end, EPA is empowered to publish and periodically revise a list of pollutants that either endanger or pose a potential risk to citizens and natural resources; however, EPA has not revised its list of criteria pollutants for many years. EPA also defines technology standards that establish what technology is required at specific emission points in specific types of industry.

Technology-specific standards are often viewed as a cornerstone of the CAA. Technology-specific standards apply to many new sources through EPA's New Source Performance Standards (NSPS) requirement, while even more stringent standards can be imposed on certain new sources in dirty-air and clean-air areas. The technology requirement for new sources in non-attainment areas is called lowest achievable emission rate (LAER) technology, while in clean air areas it is called best available control technology (BACT). Sometimes BACT and LAER end up being the same. Where that is not the case, LAER will be more stringent than either the BACT or NSPS requirements. For existing sources in dirty-air areas, regulators apply reasonably available control technology (RACT).

It should be noted that a significant conflict exists between those who support the design standard approach to regulation embodied in the NSPS, LAER, BACT, and RACT process and those who support performance standard approaches such as marketable permits or emissions fees. While the CAA in many instances requires the former, it now encourages the latter.

Criteria Pollutants and Air Quality Standards

After listing a pollutant, EPA must publish criteria to assess its effects on human health and the levels at which harm may occur. The criteria must describe the atmospheric variables and combinations with other pollutants that may alter the effects of the pollutant. EPA then issues National Ambient Air Quality Standards (NAAQSs) for each pollutant based on these criteria. Historically the NAAQSs have driven most air quality activities at the state and local levels, since most states must develop plans that demonstrate how the NAAQSs will be achieved.

Today there are six designated "criteria pollutants" (SO_2, PM10, carbon monoxide [CO], ozone [O_3], NO_x and lead [Pb]), with ambient air quality standards established for each (table 2-1). The CAA also regulates "non-

TABLE 2-1
NATIONAL AMBIENT AIR QUALITY STANDARDS

Criteria pollutant[a]	Standard type	Concentration		Averaging period or method	Allowable exceedances[b]
		Micrograms per cubic meter	Parts per million		
SO_2	Primary	80	0.03	Annual arithmetic mean	—
	Primary	365	0.14	Maximum 24-hour concentration	Once per year
SO_2	Secondary	1,300	0.5	Maximum 3-hour concentration	Once per year
PM10[2]	Primary and secondary	150	—	24-hour average	One day per year
	Primary and secondary	50	—	Annual arithmetic mean	—
CO	Primary	10,000	9	8-hour average	Once per year
	Primary	40,000	35	1-hour average	Once per year
Ozone	Primary and secondary	235	0.12	Maximum hourly average	Once per year
NO_2	Primary and secondary	100	0.053	Annual arithmetic mean	—
Lead	Primary and secondary	1.5		Maximum arithmetic mean measured over a calendar quarter	—

[a] SO_2—sulfur dioxide; PM10[2]—particulate matter, with particles with an aerodynamic diameter of 10 microns or less; CO—carbon oxide; and NO_2—nitrogen dioxide.
[b] Allowable exceedances may actually be an average value over a multi-year period.
— Nothing
Source: Federal Regulation 40 CFR 50.4-50.12.

criteria" pollutants and toxic pollutants. Federal guidelines and state regulations specify standards for emissions of criteria and many non-criteria pollutants from major and minor sources.

EPA's NAAQSs specify the primary and secondary standards—maximum allowable ambient concentrations of pollutants—that will result in desirable air quality. The intent of the primary standard is to protect public health with an adequate margin of safety. That of the secondary standard is to protect the public welfare from the adverse effects of pollutants. When establishing the primary standards, EPA is not required to consider the costs of technology or compliance, an important point. Secondary standards specifically include such factors as economic interests, impact on vegetation and visibility. EPA has established primary and secondary NAAQSs for the six criteria pollutants, as shown in table 2-1.

After passage of the CAA in 1970, EPA and the states jointly decided to subdivide the United States into Air Quality Control Regions (AQCRs). Since the CAA is administered in a delegated manner, the individual states and the AQCRs are the focal points for most air pollution control. If a particular area's air quality is worse than the NAAQSs, it is designated a non-attainment area. Areas where the NAAQSs have been met are called "prevention of significant deterioration" (PSD) areas. The classification is subject to EPA approval. An AQCR may be classified non-attainment for one pollutant and attainment or unclassified for another. In 1993 there were more than 200 inter- or intra-state AQCRs. About one-half the AQCRs were non-attainment areas for at least one pollutant. Area designations are subject to review and change over time.

Pollutants come from three general sources—mobile, area, and stationary. Stationary sources are divided into three classes: new sources, major modifications, and existing sources. The CAA defines a stationary source as any building, structure, facility, or installation that emits or may emit any air pollutant. In PSD areas, a stationary source is any grouping of pollutant-emitting activities on contiguous properties under the control of a single entity. Since states and localities can develop regulatory programs more stringent than EPA's, the definition of a source can also be more stringent.

Emissions Controls

Regulatory Reviews

The CAA provides many opportunities to impose obligations for emission controls on both new and existing sources of criteria pollutants. It mandates that before any new source is built, there will be a review of the impacts on air quality that will result from the project. Just as with the word "source," the words "new source" or "major modification" can be defined loosely or restrictively. Depending on how strict the definition is, there can be many or few "new sources."

New Source Reviews. New sources and major modifications that occur in non-attainment areas must meet specific new source review (NSR) requirements. New sources and major modifications that site in PSD areas must meet different regulatory requirements. Existing sources in non-attainment areas must meet still different regulatory requirements.

The NSR process imposes conditions that must be met before a new source (including major modifications) can get a permit to construct in a non-attainment area. These conditions include:

- The new source must install stringent emission control technology, specified as LAER.

- Owners of new sources must ensure regulators that all their other facilities in the state currently comply with emission limits and standards under the CAA.

- For a new source to be built, it must secure extra emission reductions (called "offsets") from another emitting source in the same geographic area so that the net effect of the new source's emissions will be negated.

- The emission offsets must result in a net improvement in air quality, which is achieved by requiring greater than one-for-one reductions; for example, the trade-off ratio in the SCAQMD is 1.2 and 1.3 to 1.

The operative words in the application of the NSR are "new source" and "major modification." By making these definitions tighter, more firms are caught by the NSR process and are required to install stringent control technologies and to provide offsets. Today, in the SCAQMD, both under federal and California laws, emitting even 1 ton per year of VOC or NO_x would trigger NSR and offsetting.

PSD Reviews. PSD reviews are done under an entirely different set of legislative and regulatory mandates. For example, new sources in PSD areas must install BACT. This technology may or may not be the same as the LAER requirement. (The definition of BACT and LAER has been the subject of countless lawsuits and policy statements; suffice it to say that defining technology is one of the major problems with a command-and-control approach to attaining environmental goals.)

Permits

The permit is a mechanism by which the states or EPA can enforce limitations on emissions to achieve and maintain the NAAQSs. There are construction and operating permits. Permits specify the removal of a percentage of a pollutant and limits on emissions, describe a particular pollution control technology, define monitoring requirements, and assure that emission control requirements are met. Basically, a permit is a contract between a company and the state or federal regulators. Because all emission limits must be federally enforceable, today all permits must relate back to the federal guidelines, even if a state issues the permit. Permits are the primary vehicle for enforcing environmental laws.

Offsets

New sources locating in a non-attainment area must, as noted, obtain offsets from existing sources in the same geographic area before they can get a permit to construct. The exact "area" is described in the state's rules for siting new sources; sometimes it is a city and other times an entire state.

The offset system was designed in such a way as to allow economic development and industrial growth while ensuring progress toward the attainment of the NAAQSs. The CAA requires that offsets more than compensate for increases in emission. As such, every offset transaction (described below) should result in a net benefit in air quality.

Offsets are emission reductions below the mandated emission level that one source has created so that a new or expanding source can negate its impact

---

BOX 2-3
OFFSETS

The original CAA mandated that all areas would be in attainment before 1980. In 1976 it was clear that parts of California would not be in attainment and that the provision in the CAA banning construction would be triggered. At that time, California was growing rapidly, with many jobs being created along with new emissions of hydrocarbons, NO_x, and other airborne pollutants. Politically, it was not feasible to ban construction and thereby cause this considerable economic machine to stall.

Through a creative interpretation of the CAA, the Regional Administrator in EPA Region IX, Paul DeFalco, created the Offset Interpretive Ruling. It stated that new sources could site in non-attainment areas such as the SCAQMD if no net increases in emissions resulted. EPA interpreted "net emission increases" to mean increases in emissions after applying the most stringent emissions control technology, which is LAER, and taking into account decreases in emissions achieved elsewhere. The decreases had to be over and above those required for other sources in the same airshed. As such, they would "offset" the increases in new emissions. By requiring a more than one-to-one trade-off ratio, each offset trade would result in less pollution.

The entire program of air credit trading in the United States began in 1976. The Offset Interpretive Ruling was subsequently introduced in the 1977 CAA and was followed in 1979 by the bubble policy. That policy encouraged the voluntary creation of emission reduction credits (ERCs) that could be used by existing sources to meet existing source control obligations. Offsetting is mandatory. Bubbling, or averaging of emissions, is voluntary. Since 1976 there have been over 2,500 offset transactions and fewer than 100 so-called bubbles.

on ambient air quality (see box 2-3). Offsets can be created by reductions in emissions from stationary, area and mobile sources. Mechanisms for reducing emissions include (1) installing additional pollution controls; (2) improving the effectiveness of existing pollution control devices; (3) changing inputs such as fuels and raw materials to produce less pollution; (4) closing a unit or facility; (5) reducing the number of operating hours or production shifts; and/or (6) reducing the rates of emissions. To ensure an offset transaction will improve air quality, the policy is that in non-attainment areas the offsets must be made at a greater than one-to-one ratio.

Offset transactions are subject to significant regulation. Rules dictate how the emission reduction is calculated and require that it be enforceable by federal and state regulatory agencies. They specify certain geographic restrictions, typically related to the location of the source needing the offsets, the attainment status of the area, and the type of pollutant involved. The emission reductions must meet minimum approval criteria established to get EPA certification as an offset (box 2-4). Finally, not all emission reductions can be used as offsets. In general, rules specify which types of sources can create offsets and the characteristics of the emission reduction (how and when the reduction was generated, and the like). Despite appearances, these federal regulations are quite general, and the states have considerable flexibility in designing their programs. As such, there is no uniformity among state offset programs.

Once certified by local authorities, an emission reduction becomes an ERC. ERCs can be placed in a formal or informal banking system where they are registered or can be purchased by another company. Note, however, that not every source that creates ERCs can trade them to any other source needing offsets. In addition, regulations and oversight can inhibit free trading and reduce market liquidity. Because offset markets are demand-driven and periodic in nature, and because the supply of offsets can be uncertain, the characteristics of ERC markets are different from those of ordinary commodity markets.

Offsets are most commonly used by sources in non-attainment areas that need mitigating emission reductions to obtain permits, but they can also be used in attainment areas where construction of some major new sources or major modification might produce emissions that would violate a PSD requirement or contribute to the violation of an air quality standard. These sources might also impair visibility or degrade air quality-related values in specially protected areas such as national parks. In such a case, construction might not be permissible without offsets. Offsets can also be used in PSD areas to (1) expand the margin for industrial growth in the area; or (2) alleviate detected excesses of an ambient air quality standard. While the law does not require that emissions by new sources covered under PSD regulations be offset, firms have done so because of their own strong interest in working with air regulators or their desire for good public relations.

The market for offsets can be determined by looking at demand and supply for offsets. Demand in a particular area is primarily a function of the rate of industrial growth and the threshold level at which major new sources trigger the offset requirements. Supply is a function of several factors:

- The number of firms able and willing to reduce emissions below the mandated levels.

BOX 2-4
CRITERIA FOR CERTIFICATION
BY THE ENVIRONMENTAL PROTECTION AGENCY

Typically, to receive certification, an ERC must have the following characteristics:

(1) *Real.* The emission reduction must be the result of a reduction in actual emission levels. Furthermore, the baseline from which the emission reduction is measured must be the lower of the source's actual and allowable emissions.

(2) *Quantifiable.* The emission reduction must be measurable or calculable using accepted procedures. According to EPA's Emissions Trading Policy Statement (ETPS), the quantification may be based on emission factors, stack tests, monitored values, operating rates and averaging times, process or production inputs, modeling of the performance of pollution control equipment, or other reasonable measurement. Generally, the same method must be used to quantify emission levels before and after the reduction in emissions.

(3) *Permanent.* The emission reduction must endure for the life of the new or modified source; it cannot be periodic or temporary. In most cases, "permanent" means the reduction cannot be temporary and will not be negated by future "related" increases in emissions. EPA defines permanent as an emission reduction that is assured for the life of the corresponding increase, whether unlimited or limited in duration. Permanence can generally be assured by requiring that changes in permits for sources or changes in applicable state regulations reflect a reduced level of allowable emissions.

(4) *Enforceable.* The agency issuing the permit and EPA must be able to enforce the emission reduction and its method of creation. Enforceable emissions limits must be incorporated into a compliance instrument—an air permit—that is legally binding and enforceable. According to the ETPS, the permit must specify applicable restrictions on a facility's hours of operation, production, or rates of input; future allowable levels or rates of emissions; enforceable test methods for determining compliance; and record-keeping and reporting requirements, including specified periods over which averages are calculated. To ensure the levels of emissions will not increase in the future (so as to negate the emission reduction that created the ERC), restrictions should be placed on future capacity use and hours of operation.

(5) *Surplus.* The emission reduction must exceed the level of reduction required by regulations and permits and must not otherwise be required by the air quality attainment plan. That is, "double counting" of emission reductions is not allowed. Only reductions below state and federally approved baselines can be considered surplus.

- The turnover of industry, which results in potential "shutdown"-based emission reductions.
- The level of emissions in the existing inventory of sources. Note that the offset supply in an area is always much less than the total number of surplus emission reductions that eligible sources can create. The reason is that some companies are not willing to create surplus emission reductions for transfer to another source for use as offsets.
- The stringency and form of the air quality regulations in an area. Restrictions can either limit the potential supply of offsets (for example, by prohibiting the use of credits resulting from plant shutdowns) or increase it (for instance, by allowing the use of non-traditional emission reductions, such as credits based on reductions in mobile source emissions).

With few exceptions, public entities such as hospitals, schools, airports, or prisons are not likely to sell offsets. However, they are usually required to get them if they trigger NSR (for example, to offset emissions from a new boiler that may be needed for a new hospital wing). The reason that public entities never sell offsets is that they have a unique financial situation that is not designed to reward individuals for contributing to the organization's "bottom line" profitability. Even where decision-makers are "cost minimizers" and wish to sell ERCs for offsetting, institutional impediments may discourage them from doing so. One impediment is that they do not know or cannot discover who in the organization is supposed to decide whether the ERC should be created or purchased. Another reason is a lack of clarity as to who should set the price for the ERCs. A third reason is where the money should go from the sale of the ERCs. Public agencies, it should be remembered, do not generally have the ability to reward entrepreneurial behavior.

Other sources of emissions such as large industrial facilities that might be subject to future offset requirements are often unwilling to sell as they are reluctant to part with surplus emission reductions they might need to support their own internal growth. In addition, the nature of environmental compliance and its relationship to the profit center of most businesses may lead companies to forgo opportunities to capitalize on the sale of offsets.

It is expected that the demand for offsets will increase in many major metropolitan areas. The reason is that the 1990 amendments to the CAA established lower offset thresholds for hydrocarbons and nitrogen oxide precursors to ozone in non-attainment areas. However, if the areas have an aging industrial base and little new industrial growth, the demand for offsets may not increase much as a result of the new CAA provisions. Except in extremely rare cases, there is virtually no speculation in the offset markets.

The price paid for a particular set of offsetting emission reductions is only one component in the total cost of acquiring offsets. Other expenses include search transaction costs and the "costs" of uncertainty and delay. Generally, offsets are priced on a per-unit basis, such as dollars per pound per day. Offsets in a particular area are generally sold consistent with the unit the local air agency uses in its air permitting program—tons per year or pounds per hour. Offsets for a particular project are typically purchased in a one-time transaction before the transfer of the offsets. What the buyer purchases is the right to emit a specific pollutant at a certain rate such as 100 tons per year in perpetuity.

Offset regulations mandate that buyers purchase, as noted, more offsets than the expected increase in emissions from a facility. A major issue in the negotiations over the price of an offset is which party will pay for the additional offsets required by the offset ratio. The seller generally insists the buyer pay the per-unit price for all offsets that are transferred from its facility. The buyer obviously prefers to pay for only those offsets it will actually apply under its permit (that is, the amount of its net increase in emissions).

Historically, the unit prices for offsets have not approached the marginal cost of controlling emissions (that is, the replacement cost of the offsets). The reason is that many offsets are created through plant shutdowns rather than through the installation of additional controls.

State Air Pollution Control Plans

Federal law mandates that each state prepare a state air pollution control plan (called a state implementation plan, or SIP) that sets forth goals and procedures for attaining, maintaining, and enforcing the NAAQSs in each AQCR. Specifically, the CAA mandates that the states attain and maintain the NAAQSs by limiting emissions of specific air pollutants at their source. The

---

BOX 2-5
REQUIRED CONTENT OF STATE PLANS

State SIPs must contain the following elements:

(1) Emission limitations and schedules for compliance.
(2) Preconstruction review, transportation controls, and air quality maintenance plans.
(3) Programs for issuing construction permits in non-attainment and attainment areas.
(4) Procedures to monitor air quality and compile and analyze data.
(5) Plans for staffing, funding, and obtaining the required legal authority to impose requirements mandated by the CAA.
(6) Inspection and testing of motor vehicles subject to the limitations on emissions from mobile sources.
(7) Plans to cover administrative costs through fees for permits from regulated sources.
(8) Processes for amending the SIP.

states must submit their plan to EPA and get its approval. The plan must demonstrate how the state will meet and protect ambient air quality (see box 2-5). The CAA specifies that EPA can reject a SIP and promulgate alternative regulations if the state presents an unacceptable plan to EPA. In the interim, EPA can ban proposed construction, fully or conditionally, in a non-attainment area until it approves the SIP. EPA can also limit funds for highway construction. Since the consequences of failing to produce a satisfactory SIP can be draconian, EPA has been reluctant to use its powers. It has, however, imposed bans on construction and taken over air regulatory functions.

Typically, any change in a facility's permit requires a corresponding revision of the SIP. The process of revising a SIP is notoriously slow and administratively burdensome. These lengthy and uncertain procedures have resulted in firms' trying to avoid NSR at all costs.

THE ADMINISTRATIVE STRUCTURE OF THE SOUTH COAST AIR QUALITY MANAGEMENT DISTRICT

At the heart of the SCAQMD's activities is the AQMP, a 20-year action plan detailing how the basin will be brought into compliance with federal and state pollution standards. Much of this plan is embedded in the SIP, although some of the plan is outside the SIP. The District works with the Southern California Association of Governments (SCAG) to update the AQMP every three years. The plan must be approved by the District's governing board and by state and federal officials.

The SCAQMD is responsible for implementing the SIP. It takes the lead in efforts to reduce emissions from industry and works with the ARB to reduce motor vehicle and fuel emissions. It is also involved in lowering emissions from commercial products such as paints, solvents, and lighter fluids. Finally, the District takes part in monitoring and regulating air emissions of toxic chemicals and in phasing out chlorofluorocarbons in the Basin.

Governance and Organizational Structure

A 12-member board of local elected and appointed officials governs the SCAQMD. An executive officer carries out day-to-day management. Nine offices report to the executive officer: chief prosecutor; district counsel; finance; technology advancement; major source compliance; planning; computer systems; technical support; and small business and consumer products (see figure 2-2 and box 2-6). In FY1992-93, the District had more than 1,000 full-time equivalent (FTE) positions authorized, of which it had funded 981.

This organizational structure was established as part of the New Directions program, which was adopted in April 1992. The goal of New Directions was to improve "customer service" when dealing with regulated industries, especially small businesses. The rationale for the changes was the large number of small businesses that required air pollution permits and that could

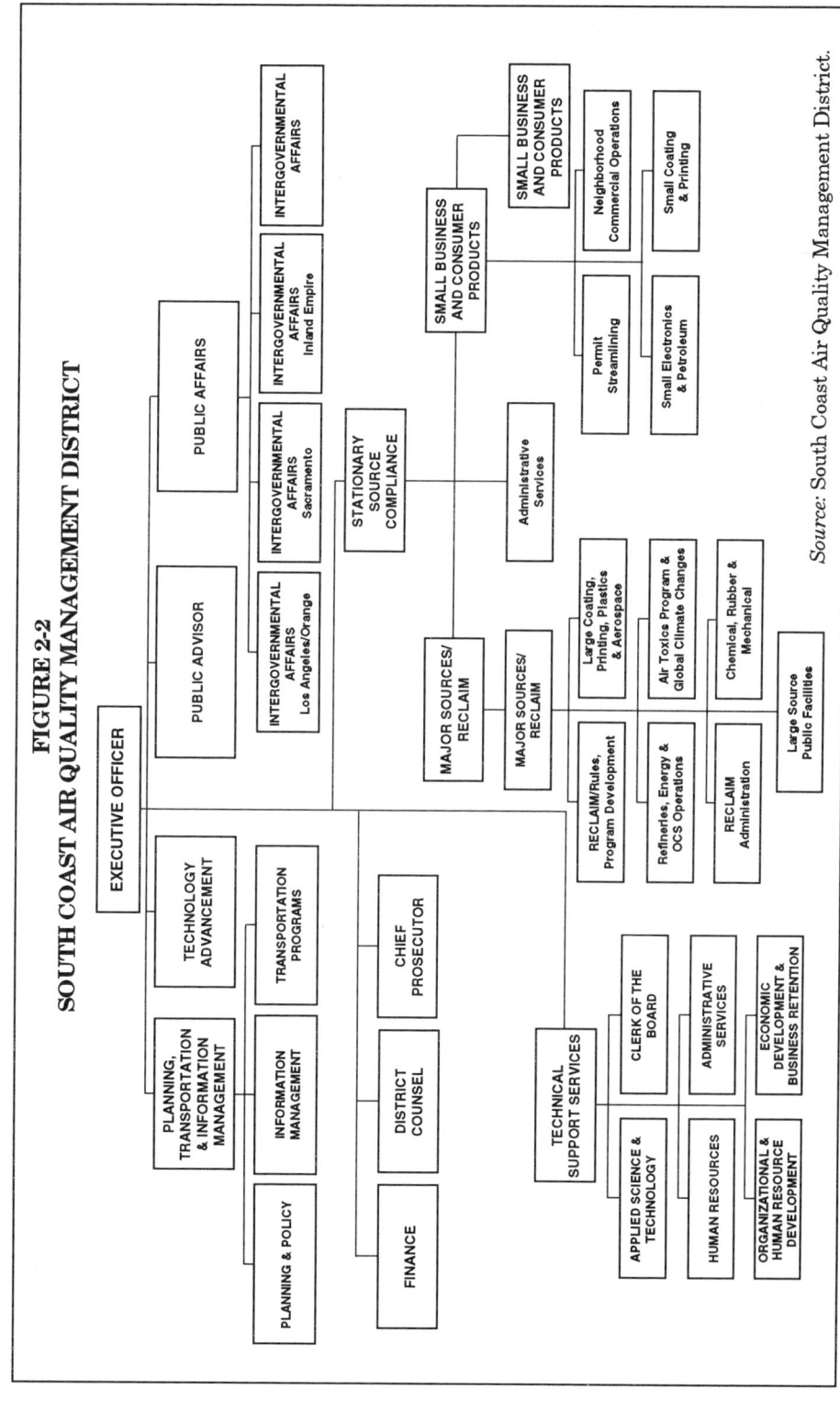

FIGURE 2-2
SOUTH COAST AIR QUALITY MANAGEMENT DISTRICT

Source: South Coast Air Quality Management District.

BOX 2-6
THE SCAQMD'S BUILDING BLOCKS

The key offices and activities of the SCAQMD are described briefly below. They illustrate how complex and comprehensive the District's mission of achieving air quality standards is. They also illustrate the wide range of staff skills, knowledge, and abilities required.

- **Chief prosecutor.** This office represents the District at variance hearings, orders for abatement, revocation of permits, and appeals. Investigators from this office focus on the effective processing of notices of violation using various administrative, civil, and criminal proceedings to resolve cases.
- **District counsel.** This office advises the board and staff on all legal matters except those related to enforcement of District rules.
- **Finance.** This division handles payroll, budgets, financial planning, cash management, and maintenance of the management information system. Budget, payroll, and program accounting are linked to the District's Work Program Tracking System, which provides detailed information on actual and planned program costs.
- **Technology advancement.** The primary function of this division is to facilitate the development of advanced technology. The division works closely with industry to research and refine air quality control technologies. Recently, it has expended considerable effort to develop flexible fuel vehicles.
- **Major source compliance.** This activity involves review and authorization of permits to construct and operate sources that emit pollutants. Each permitted source is subsequently subject to inspection and annual permit review and develops rules related to major sources.
- **Planning.** This activity ensures that planning results in practical and implementable rules. It carries out air quality, point source and economic modelling and manages the emissions inventory. This group also develops and submits the region's SIP.
- **Computer systems.** This activity includes operation of the Enhanced Automated Equipment Inventory System that enables the District to manage its permit processing and compliance operations. In addition, new applications are created to upgrade the support of permit processing, laboratory analysis, and air quality modeling.
- **Technical support.** This activity involves operation of the District's air monitoring stations, carries out specialized ambient monitoring, conducts source tests, and operates the District's laboratories.
- **Small business and consumer products.** This division reviews and authorizes permits to construct and operate sources classified as small sources (non-RECLAIM sources). It is also responsible for the implementation of programs related to consumer products.

not dedicate expert staff to dealing with the District. The New Directions program began with a reorganization of the District's permit and compliance activities to better meet the needs of end-users. The Stationary Source Compliance division resulted from the merger of the former Rules and Engineering Divisions. The new division is divided into three branches—Major Sources/RECLAIM, Small Business and Consumer Products, and Administrative Services. Major Sources/RECLAIM and Small Business and Consumer Products in turn have sub-branches dedicated to particular industrial sectors and provide expertise in rules, engineering, and permits. The two branches are empowered to answer all questions a firm has about permits without having to refer to other divisions or higher authorities. This reorganization resulted in a matrix approach to internal decision-making on permits and compliance. All District employees have received training in customer service. The SCAQMD provides industry with opportunities to evaluate the services received.

Other reforms include courses on how to write permit applications and on-site assistance with compliance.[2] The SCAQMD also instituted an important temporary measure that has proved effective: it designated part of its headquarters' basement as a "dug-out" where the backlog of permit applications was tackled. By 1992, that backlog had reached several thousand applications, some of which required as little as two hours to process. Various branches would rotate into the dug-out. By the fall of 1993, the backlog had been eliminated, and the dug-out is now being used to prepare and send out RECLAIM permits.

Finally, the New Directions reorganization was used to group the most customer-oriented and innovative staff to key positions in anticipation of the development of the new regulatory program, RECLAIM. Top personnel involved in planning and rule-making were selected to carry out the feasibility study for RECLAIM. They then went on to design the program.

Revenues and Expenditures

Since its establishment in 1978, SCAQMD'S primary revenue sources have been the emission and permit fees charged to polluters based on the emission of contaminants, toxic compounds, and ozone depleters. The primary expenditures have been salary and employee benefits and professional and special services.[3] Over time, as enforcement increases, emissions are expected to decline and, with them, the revenues from emission fees. While revenues are expected to stabilize in conjunction with the implementation of RECLAIM, operating the program will require greater expenditures on such items as salaries and expenses, utilities, and growth of the Air Quality Assistance Fund (a trust fund established by Assembly Bill 2444, which requires that the

---

[2] An interesting outcome of the New Directions program is that the amount of fines the District has been collecting has decreased.

[3] "Three-Year Budget Forecast Fiscal Years 1993-94, 1994-95, and 1995-96," South Coast Air Quality Management District, Diamond Bar, California, pp. 1-10.

District allocate at least $1 million annually from funds it receives from civil penalties to provide loan guarantees for small businesses installing pollution control equipment).[4]

The budgetary picture has been complicated by recent state legislation that sets a cap on the District's overall budget equal to the revenue generated during FY1993-94. This cap will affect future programs, although exactly how is still unclear. The revenue shortfalls noted above have already led the District to initiate a staff reduction program; a layoff plan that will affect 127 employees has been announced. The program will be initiated in the spring, to be completed in time for the next fiscal year.

Revenues

The SCAQMD is required to establish regulations for the collection of fees to support its operations. The SCAQMD derives its revenues from several sources, including emission fees, fees from permits, and operating fees. For example, while the District's roles and responsibilities are established under federal and state statutes, it gets 6 percent of its revenues from government financial assistance, mainly in the form of grants. In 1990, the first three sources provided more than 84 percent of the District's revenue, with emission fees accounting for 40 percent, operating fees 28 percent, and permit fees 16 percent.[5]

In the past, these sources of revenue were sufficient to enable the District to carry out its mission. During the late 1980s, the strong economic growth in the Basin led to rapid growth in the SCAQMD, which expanded to keep pace with the large increases in permit applications and related revenue and in emission fees. In the early 1990s, the picture changed. In 1991 and 1992, the economic recession led to a contraction in the District's budget, as new applications for permits fell off and a drop in output curtailed fees from emissions. In 1991, for example, there were no increases in the revenues from emission fees, although they did rise 6 percent in 1992. Similarly, the economic recession has kept the District from increasing its fees to counterbalance the shortfall. Based on a three-year budget forecast covering FY1993-94, 1994-95, 1995-96, and 1996-97 (table 2-2), it is expected that the emission fee revenue base will contract somewhat. Annual revenues are expected to stay the same. Processing fees are also expected to remain stable as RECLAIM and AQMP control measures are implemented. Revenues from mobile sources and subscriptions are expected to increase slightly, while revenues from toxic hot spots are expected to decrease slightly.[6] It is clear the District can no longer rely on emission fees and other traditional sources to support the budget needed to carry out its mission to the degree it has in the past.

---

[4] "Draft Budget Fiscal Year 1992-93," South Coast Air Quality Management District, May 1992, Diamond Bar, California, p. 13.

[5] "Report to the South Coast Air Quality Management District for the Fee Assessment Study," KPMG Peat Marwick, Los Angeles, March 1990, p. II-6.

TABLE 2-2
THREE-YEAR REVENUE FORECAST BY MAJOR ACCOUNT
(dollars)

REVENUE TYPE	FY 1993-94 Base	FY 1994-95 Forecast	FY 1995-96 Forecast	FY 1996-97 Forecast
Emission fees	$25,750,000	$24,586,400	$24,136,100	$23,828,500
Annual operating fees	27,604,000	25,782,700	27,761,300	28,793,100
Permit fees	9,636,000	8,332,300	8,901,700	9,572,000
Air Resources Board subvention	3,066,000	3,066,000	3,066,000	3,066,000
U.S. Environmental Protection Agency grant	3,000,000	3,000,000	3,000,000	3,000,000
Mobile sources	21,676,000	21,700,000	23,934,000	24,652,000
Transportation programs	1,900,000	2,391,400	2,750,500	2,571,400
Toxic hot spots	3,900,000	3,989,700	3,717,600	3,421,300
All other	5,877,700	4,951,500	4,032,800	3,795,700
Total revenues	$102,409,700	$97,800,000	$101,300,000	$102,700,000

Source: "Three-Year Budget Forecast Fiscal Years 1993-94, 1994-95, and 1995-96," South Coast Air Quality Management District, Diamond Bar, California, p. 2.

The SCAQMD's financial planners have been contending with another problem: it is difficult to calculate precisely how much revenue will be received from the emissions fees and penalties until it is collected. Most of the emissions fees arrive in March, at which time the finance division is overloaded with associated paperwork. Although the penalty fees arrive throughout the year, they account for less than 3 percent of the budget, and even that share will decline under the New Directions program.

One possible answer to this budget dilemma is to base the fees on proportional rather than actual emissions, that is, on the proportion of the Basin's total emissions each firm is entitled to, rather than on the actual emissions it generates. Proportional emissions remain constant, even as total emissions decline, so that revenues from this source would not change over time. In short, even as a firm reduced its emissions, its fees might stay the same or increase (a result that might be counterintuitive to those who pay the fee). Firms could, however, lower their proportional emissions fee by trading ERCs. The District is authorized to collect fees on proportional emissions.

The population of firms that pays fees is heterogeneous. As of 1992, of the 30,000 firms in the Basin that had received permits, half had only one permit on file and two-thirds had two or fewer. Half the firms paid less than $195 a year, 75 percent less than $500 a year. About 25 firms paid the bulk of the emission fees the District collected.

Expenditures

The SCAQMD's expenditures fall into the following broad categories (the share of each in total expenditures appears in parentheses): stationary source compliance (30.7 percent); technical support services (26.2 percent); general expenditures (14.4 percent); planning and technology advancement (12.7 percent); and transportation and public affairs (8.7 percent) (table 2-3). In terms of use, 61 percent of the budget goes for salaries and benefits, 38 percent for services and supplies, and the remainder for fixed assets.

Based on a three-year budget covering FY1993-94, 1994-95 and 1995-96, it is expected that expenditures will increase. Salaries and employee benefits should rise slightly, enforcement costs will grow a little, and contributions to the Air Quality Assistance Fund will add $1 million during 1993-94. Other expenditures such as technology advancement, rules, and technical activities are expected to remain constant or increase slightly.[7]

Staffing

The following discussion is presented to illustrate the mixture of skills and capabilities needed to design and implement the District's regulatory initiatives. The District relies on highly skilled, energetic staff in virtually every

---

[6] Op. cit., "Three Year Budget Forecast...."
[7] Op. cit., "Three Year Budget Forecast...."

TABLE 2-3
THREE-YEAR PROGRAM FORECAST BY CATEGORY
(dollars)

PROGRAM CATEGORIES	FY 1993-94 Base	FY 1994-95 Forecast	FY 1995-96 Forecast	FY 1996-97 Forecast
Enforcement	$23,685,105	$16,747,459	$17,273,875	$17,703,148
Episodes	387,031	237,057	247,839	254,079
Interagency	1,361,298	1,281,109	1,306,026	1,364,494
Operational support	22,252,185	21,787,916	22,730,654	23,158,937
Outreach	6,109,835	5,441,793	5,648,731	5,754,507
Permits	17,758,357	19,705,709	19,703,707	20,069,123
Planning	6,922,762	6,788,756	6,690,187	6,492,449
Rules	6,871,913	5,919,395	6,059,768	6,067,656
Technology advancement	7,510,161	7,768,110	8,832,758	8,864,267
Technical	10,532,403	9,135,628	9,719,111	9,821,365
Transportation	3,186,228	2,975,693	3,072,551	3,151,609
Total expenditures	$106,577,278	$97,788,625	$101,285,207	$102,701,634

Source: "Three-Year Budget Forecast Fiscal Years 1993-94, 1994-95, and 1995-96," South Coast Air Quality Management District, Diamond Bar, California, p. 2.

aspect of its operations. It emphasizes hiring people with advanced degrees and employs many individuals with Ph.D. and master's degrees in a variety of technical and management positions. In addition to the technical positions in air modelling and economic forecasting, Ph.D.s are involved in program management and other management tasks. The District generally recruits directly from colleges and universities. As a result, its workforce is young.

During recruitment, applicants are asked a series of "situational questions" that test their ability to generate innovative approaches within well-defined parameters. Creative employees, managers assert, thrive in the low-control, high-reward and, entrepreneurial environment of the SCAQMD.

Senior management is recruited largely from the outside rather than promoted internally—well over half the top dozen managers did not start their careers at the District. The staff is well-diversified in terms of gender, race, and ethnicity.

Career potential at the District has two phases. During the first phase, the lower-level employees are offered many opportunities for incremental advancement. The potential for career growth shrinks during the second phase. Unfortunately, the recent budget contraction is delaying the advancement, of many employees who had grown accustomed to a fast pace of promotion. Staff who came on board in the early 1980s benefitted from the rapid growth of the organization and now hold positions of responsibility. Those who joined the District more recently will find their career growth potential slowed (the result of the budget cuts forced by the recession). Still, they are unlikely to fall into a tedium trap.

The retention rate at SCAQMD has been high. Staff members say they work for the SCAQMD because of the competitive salaries and benefits package, the opportunity for rapid career advancement, and the openness and responsiveness of the public service aspect of the work. They also like the District's reputation as an innovative regulatory body.

Some staff, however, do not feel the performance evaluation system rates employees accurately. They question the integrity of the review system, in that it does not track their performance against explicit goals and objectives. Some staff perceive that managers reward and recognize employees according to personality traits such as intelligence, initiative, and flexibility, rather than for high standards of performance. In fact, mobilization of the vision, energy, and resources of the employees is, to a considerable extent, linked to the personality traits of employees and the nature of the work.

Salary information is public. All but senior staff receive 1.5 times salary for overtime hours. Senior staff earn salaries in the range of $70,000 to $110,000 a year and are eligible for annual performance awards as high as 8 percent of total salary. Anyone who does not receive a performance award usually leaves the organization. Non-monetary benefits such as parking spaces, employees' cafeteria, and social events are available to all employees. In general, employees who appear to have high career growth potential are tapped for innovative work such as RECLAIM, which is viewed as a fast-track assignment that increases promotional opportunities. Staff currently involved with RECLAIM anticipate their next assignment will be to develop another innovative program and gauge their potential for professional growth in terms of designing and developing, rather than operating, programs.

Major Program Categories

As described, the District's mission is to reduce air pollution in the Basin to legislated levels. To carry out that mission it performs a number of functions that can be divided into the following program categories:

- **Enforcement**, which deals with stationary sources. Activities include inspections of regulated facilities, issuance of notices of violation and other injunctions, responses to citizen complaints, support to other agencies, and training and outreach programs.

- **Episodes**, which refers to alerting the public through computer networks and radio broadcasting when the level of air pollution poses a major health threat; conducting inspections to determine compliance with the episode abatement plan; providing technical assistance to fire, police, and health agencies; and monitoring data to update and verify air quality models.

- **Interagency**, which refers to interaction with local, state, and federal agencies on issues of policy, legislation, jurisdiction, and joint ventures.

- **Outreach**, which involves a wide variety of public information campaigns and outreach programs.

- **Permits**, which entails the approval or denial of applications for permits to undertake construction and operate equipment that might have an impact on air pollution levels.

- **Planning**, which involves development of the AQMP, including analysis of air quality data, estimation of pollutant emissions, development of control strategies, and projection of future pollution levels and socioeconomic impacts.

- **Rules**, which entails the assessment of control technologies and costs, solicitation of public input, and the development of the administrative rules for the permit program.

- **Technical support**, which consists of ambient air quality data collection and analysis, source testing, and laboratory services, as well as evaluation of the potential impact of proposed pollution control strategies on the regional economy and environment.

- **Technology advancement**, which involves the development of clean fuels and low emission technologies by continually assessing the status

of research, development, and demonstration activities, as well as working with industry and public agencies.
- **Transportation**, which consists of a wide variety of transportation control programs, such as ride-sharing, and promotion of low-emitting vehicles, as well as technical support to local governments in developing programs to manage congestion.
- **Administration**, which involves management of facilities, budgeting and finance, human resources, support for the Governing Board and the Hearing Board, and information management services.

Organizational Culture

The District's culture has two primary characteristics that are sometimes in conflict—it is both innovative and inward-looking. The SCAQMD has a tradition of innovation, including in the areas of developing regulations to govern emissions from landfills, new control technologies and strategies, computer networking, ride-sharing, and scrapping of old cars. Innovation is emphasized during recruitment, management leadership training, and selection for career advancement. A general awareness that the District is being scrutinized by regulatory agencies throughout the nation and world sustains innovation. Employees and managers see themselves as trail-blazers and refer proudly to their cutting-edge atmospheric and economic modelling teams, to a series of regulatory innovations, and to support for scientific innovations, including first-of-a-kind fuel cells and hook-ups for electric vehicles.

Despite this culture, some staff have resisted new programs. For example, some involved with the enforcement of rules appear to have opposed RECLAIM. In the past, employees who shunned anything short of a sure thing were not a problem, since the District was growing quickly enough that supportive employees could be promoted diagonally past those resisting change. More recently, that attitude became a problem that had to be addressed through the reorganization under the New Directions program.

As to the inward-looking nature of the culture, it appears that both external and internal factors cause employees to look within the District for solutions to problems. The combination of close scrutiny from outside, expansive regulatory and punitive powers, independent sources of revenue, relative geographic isolation of the headquarters building, and creative talents of District staff has contributed to a myopic sense that the District's actions and decisions are necessarily correct. Outside criticism and suggestions of alternative methodologies appear to have little impact until problems reach a crisis. The most obvious example is the backlog of permit applications. The District did not formally address the foot-dragging until ongoing complaints from regulated sources were accentuated by the economic recession and job losses. While the District is not alone in lying back and responding only after the house catches fire, reportedly its insular view of problems and occasional tendency to ignore the views of outsiders have resulted in some alienation of environmentalists and regulated parties. A strong belief in the District's prerogatives and the assumption by some managers that they have the sole right to decide how their core tasks are carried out have, at times, resulted in a we-versus-they attitude. It is worth noting that several interviewees observed that while command-and-control systems tend to promote an insular view, market-based programs require that the public and private sectors be engaged.

Leadership

As noted, day-to-day management is the responsibility of the executive officer, who also handles legislative liaison and represents the SCAQMD before other governmental bodies. On many major decisions, the executive officer consults an executive council composed of the senior managers.

The executive officer exerts considerable control over the District's management hierarchy. The jobs of the top 20 officers are not protected—they sign contracts that commit them to serve at the pleasure of the executive officer. The executive officer recruited the majority of the top officers from outside the SCAQMD, either from industry or other government agencies.

Three effects of the District's leadership bear mentioning. Since the employment status of senior managers is somewhat precarious, they may tend to be overly optimistic in reporting results from their divisions to the executive officer and may gloss over the negative aspects of their divisions' activities. Second, while the executive officer is able to change directions rapidly, that capacity may also mean a lack of continuity and stability. Third, the size of the organization (which has doubled since the mid-1980s) may make it more difficult for one individual to exercise authority effectively in so many different areas. For example, transportation regulation is quite different from stationary source regulation, but equally demanding. The demands of planning and designing new programs may compete with the demands of operating existing programs.

THE SCAQMD BANKING AND TRADING MARKET

As noted, the CAA and associated policies developed by EPA allow companies to create ERCs, store them, and use them in permitting activities. These ERCs can be employed to meet or avoid certain air permitting requirements. For example, a new source of non-attainment pollution siting in a non-attainment area must obtain ERCs to offset its new emissions. In addition, existing firms can rearrange their emissions control strategy by creating extra ERCs at one emission point in lieu of reductions at another point (such an offset inside an existing facility is called a bubble). Alternatively, a company can use an ERC in a process called netting (internal facility-wide offsets), which allows the firm to avoid certain regulatory requirements.

EPA has codified the process of creating, storing, and using ERCs in its

ETPS and subsequent Economic Incentive Rules. While the ETPS was developed over an eight-year period, the rules governing economic incentives were developed in three years, to be finalized by the end of 1993. These procedures state how firms create ERCs; how ERCs can be protected through the development of administrative procedures called emission banks; and how ERCs can be used in offsets, bubbles, and netting.

Because of a progressive regulatory philosophy, severe air quality problems, and rapid economic growth during the 1970s and 1980s, the SCAQMD has been the leader in emissions banking and trading. As noted, the federal government mandates that the SCAQMD control air pollution from new and modified facilities consistent with its NSR program. Essentially, the District has been using a traditional command-and-control approach to regulate air pollution. The rules established for the District's NSR program are described in District Regulation XIII (see box 2-7). The two key activities the District undertakes in implementing the NSR are the issuance of air pollution permits to industries in the Basin and verification that the stipulations in the permits are followed. These activities establish the baseline for calculating the offsets demanded by new sources and the ERCs created when existing sources shut down or modify their facilities, equipment, or processes. The permit must stipulate how a piece of equipment is used and what controls apply to it. Each unit with the potential to emit above certain levels must apply for and receive a permit to construct and operate. Regulated firms have to submit annual reports on their emissions and are subject to inspections to assure compliance with the stipulations in the permits.

The SCAQMD's Offset Program

The SCAQMD has the largest offset program in the United States. It has revised its regulations several times since October 5, 1979, when it adopted its

---

BOX 2-7
SELECTED PROVISIONS OF DISTRICT REGULATION XIII

(1) All new and modified facilities, regardless of size, must offset increased emissions.

(2) Offsets must be obtained from facilities that are generally upwind from or in the same area as the increased emissions.

(3) ERCs, the source of the offsets, are calculated as if the equipment were already BACT.

(4) The District runs a community emissions offset bank from which low-polluting facilities can obtain offsets free of charge. The District also, as noted, set up a separate allocation called the priority reserve from which essential public services, such as schools and hospitals, can obtain offsets.

first emissions banking and trading regulations. It adopted its most recent NSR-specific revision on June 28, 1990, with the changes to take effect October 1, 1990. The revisions were necessary because the existing regulations were not perceived to allow adequately for economic growth while ensuring that new and modified facilities did not interfere with progress in meeting the national standards for ambient air quality. A driver for change was the gaming of the system by companies that chose to stay below thresholds or made few emissions adjustments. At the threshold, they would internally offset. Another factor was that minor sources did not need to offset. On October 15, 1993, NSR was again modified as part of the RECLAIM rules.

One reason the District has been the site for active offset trading is the existence of the emissions bank. The emissions bank is nothing more than administrative procedures for creating and storing ERCs. These credits can be sold or applied to external offset trades between companies, or can be used as netting. While only a few air districts have taken the time and spent the money to develop an emissions bank, a well-run bank is critical to the success of an air credit trading program. Table 2-4 reviews trading activity in the District from 1985 through 1992.

The evolution of market-based air pollution programs in the United States started in 1976 with EPA's adoption of the offset policy, which, as noted, applied only to new sources of criteria pollutants. In 1977 this policy was incorporated into the CAA. Later, air credit trading was extended to existing sources of criteria pollutants. While many economists and policy advocates pushed for a more comprehensive approach to using economic incentives for reaching air pollution control goals, it was not until 1989 and 1990 that more-comprehensive approaches were developed.

In 1990 Congress amended the CAA and specified a RECLAIM-like system for controlling sulfur precursors to acid rain. These provisions are described under Title IV of the CAA. At the same time, the District was beginning its deliberations on how to extend air credit trading into a more comprehensive approach to attaining ambient ozone standards. Today, both the Title IV and RECLAIM programs stand as examples of how the 1976 offset policy has been successfully expanded to provide more-powerful regulatory tools.

The SCAQMD emissions bank currently holds more than 400 certificates for ERCs, probably the majority of the banked ERCs in the United States. The certificates are numbered and show the owner, type of pollutant, quantity of credits, and trading region in which the source creating the pollution is located. If appropriate, the bank registry also indicates any previous certificate numbers under which the certificate was issued so that all trades can be tracked. The emissions bank does not distinguish between credits created through changes in processes and shutdowns, as today they are treated the same in terms of banking and trading.

While the offset trading program is conceptually simple, the rules for trading offsets are very detailed. The SCAQMD requires that any increase in emissions must be completely offset (a zero threshold level). This means that

TABLE 2-4
OFFSET TRADING IN THE SOUTH COAST AIR QUALITY MANAGEMENT DISTRICT, 1985-92

A. VOLATILE ORGANIC COMPOUNDS (VOCs)

Year	Number of offset trades represented	Total qty traded		Average price	
		Lbs/day	Tons/yr	$/lb/day	$/ton/yr
1985	1	197	26	343	2,639
1986	6	1,097	142	495	3,807
1987	5	1,613	210	473	3,637
1988	8	1,608	209	585	4,500
1989	4	957	124	1,014	7,798
1990	3	635	83	683	5,256
1991	8	3,915	509	1,723	13,253
1992	1	148	19	1,200	9,231

B. NITROGEN OXIDE (NO_x)

Year	Number of offset trades represented	Total qty traded		Average price	
		Lbs/day	Tons/yr	$/lb/day	$/ton/yr
1985	1	1,290	164	913	7,019
1986	8	4,079	530	730	5,613
1987	0	0	0	n.a.	n.a.
1988	3	297	39	473	3,637
1989	1	800	104	548	4,212
1990	2	190	25	2,200	16,923
1991	14	189	25	2,394	18,413
1992	0	0	0	n.a.	n.a.

C. PARTICULATE MATTER (PM)

Year	Number of offset trades represented	Total qty traded		Average price	
		Lbs/day	Tons/yr	$/lb/day	$/ton/yr
1985	0	0	0	n.a.	n.a.
1986	3	424	55	474	3,642
1987	0	0	0	n.a.	n.a.
1988	0	0	0	n.a.	n.a.
1989	0	0	0	n.a.	n.a.
1990	0	0	0	n.a.	n.a.
1991	1	20	3	4,000	30,769
1992	1	6	0.8	3,000	23,077

D. SULFUR OXIDE (SO_x)

Year	Number of offset trades represented	Total qty traded		Average price	
		Lbs/day	Tons/yr	$/lb/day	$/ton/yr
1985	1	1,369	178	548	4,212
1986	4	1,569	204	540	4,154
1987	0	0	0	n.a.	n.a.
1988	0	0	0	n.a.	n.a.
1989	0	0	0	n.a.	n.a.
1990	0	0	0	n.a.	n.a.
1991	1	20	3	a	a
1992	0	0	0	n.a.	n.a.

E. CARBON OXIDE (CO)

Year	Number of offset trades represented	Total qty traded		Average price	
		Lbs/day	Tons/yr	$/lb/day	$/ton/yr
1985	0	0	0	n.a.	n.a.
1986	4	60	8	556	4,276
1987	0	0	0	n.a.	n.a.
1988	0	0	0	n.a.	n.a.
1989	0	0	0	n.a.	n.a.
1990	0	0	0	n.a.	n.a.
1991	1	4	0.5	675	5,192
1992	0	0	0	n.a.	n.a.

F. TOTAL OFFSET TRADING

Year	Number of offset trades represented	Total qty traded		Average price	
		Lbs/day	Tons/yr	$/lb/day	$/ton/yr
1985	3	2,826	367	601	4,623
1986	25	7,229	940	584	4,492
1987	5	1,613	210	473	3,638
1988	11	1,905	248	554	4,262
1989	5	1,757	228	921	7,085
1990	5	825	107	1,290	9,923
1991	15	4,148	539	2,002	15,400
1992	2	154	20	2,100	16,154

n.a.–Not available.
[a]This quantity represents one trade between two parties for which no payment was made.
Source: South Coast Air Quality Management District.

all increases in emissions must be negated. In practice, the offset ratio in the SCAQMD has been preset at 1.2:1 for all sources except those using the community bank or priority reserve (see below), for which the ratio is 1:1. The ratio for internal trades is also 1:1. The majority of trades are internal. The source acquiring the offset must get ERCs from another source in one of the 38 trading regions as specified by the SCAQMD. There are specific trading rules governing the disposition of trades from each region to the other 37 regions so that wind patterns and potential impacts are accounted for in each offset transaction.

As part of its program, the SCAQMD has established a community bank and a priority reserve for offsets. The community bank is for sources that do not own any ERCs and need just a small quantity—less than 30 pounds (lbs) a day for VOC, 40 for NO_x, 60 for SO_x, 30 for PM_{10} and 220 for CO. According to an official at the District, the regulation may be changed to allow only small businesses—defined as companies with less than 10 lbs. a day of total emissions and less than $0.5 million in revenues—to use the community bank. Under existing regulations, more offsets have been traded through the community bank than through the third-party market (two sources trading directly). In 1991, 15,346 lbs a day of offsets went through the community bank, compared with only about 4,500 lbs a day through the third-party market. Either only small sources are expanding and locating in the SCAQMD, or the incentives are so great for sources to keep emissions below the thresholds that they will go to great lengths to be eligible for the community bank. As to the priority reserve, it provides essential public services with ERCs free of charge.

Even though the SCAQMD regulations are fairly easy to follow and specify the procedures for calculating emission reductions, some features discourage sources from participating in the emissions banking system. One is that the restrictiveness of the rules makes it extremely difficult to create bankable credits by any means other than closing a plant. Given how strict the District's rules are, getting more emission reductions by over-controlling emissions is difficult. Another reason is that the fee the District charges for processing ERCs is high. It may be particularly high for companies with multiple permits (the SCAQMD issues separate permits for individual pieces of equipment). Under RECLAIM, all equipment permits at a single facility will be combined into a single facility permit. The SCAQMD can influence the costs of transactions by slowing or speeding up approval, by requiring significant or just minimum documentation for each trade, or by fast tracking or slowing the administrative review by placing other activities ahead of trading actions.

A further disincentive to participate in RECLAIM is that the SCAQMD applies stringent discounts to actual emission reductions to account for the further reductions a facility will have to make to meet future emission targets, to meet the requirements of proposed future rules, and to achieve decreases in emissions specified in the AQMP and required as if BACT-level controls were applied to the emissions. The discounts may be sufficiently large that a facility cannot justify the effort or cost of submitting a banking application. Uncertainty over the level of the discounts that will be applied is another inhibiting factor. Industry is also frustrated by the length of time the District takes to process an ERC application. In some cases it has taken more than a year to process an application that required only a couple of hours to complete. The SCAQMD's excuse is that applications for permits take precedence over applications for ERCs, and the District had a backlog of the former. As noted, the SCAQMD has addressed this problem.

The benefits of the emissions trading system have been substantial. Without offsetting, new sources would not be able to site in the District and existing companies could not expand. At the same time, several industries have threatened to leave the District or have already left the area because they contend the SCAQMD makes operations too expensive and burdensome. Few industries have actually attributed their movement from the District just to expensive and burdensome operations. The truth is that the cost of operating within the SCAQMD is not significantly related to the offset program. Offsets account for less than 1 percent of the life-cycle cost of a new facility when capital, labor, taxes, and other expenditures are properly accounted for.

Unfortunately, many people have carried their criticism further, saying the entire regulatory program has contaminated the business climate in the SCAQMD. As a result, the trading program, which is popular, has been branded with the same negatives that characterize the traditional command-and-control approaches.

THE LATEST SCAQMD INITIATIVE: RECLAIM

The SCAQMD is the only extreme non-attainment area in the United States for ozone. This air pollution problem is serious and ubiquitous throughout the District. Developing rules on a source-by-source basis has become both expensive and time-consuming. In addition, as regulators strive for more and more emission controls, they are forced to be more and more intrusive in the day-to-day activities of even the smallest firms. SCAQMD regulators have continually communicated with industry and environmental communities, and partly out of those relationships, a common understanding grew that the existing system of pollution control was failing to meet statutory requirements.

The District could not take the time (one to three years to develop a regulation for a source category such as industrial boilers or degreasing operations), nor could the District afford the costs in staff and contractors. Regulations of existing sources of non-attainment emissions thus had to change.

In response, regulators and others looked for new ways to regulate both existing and new sources. The result is that the SCAQMD is implementing an alternative regulatory program, RECLAIM, to meet the Basin's NO_x and SO_x air quality objectives. RECLAIM is a significant departure from the emission reduction strategy the SCAQMD had been pursuing to attain the air quality standards for VOC, NO_x, and SO_x emissions from stationary sources. It has been designed to reduce those emissions from sources to the same extent they would have been under existing and future command-and-control regulations while affording sources maximum flexibility in deciding how best to control those pollutants and stimulating innovation and technological advances. RECLAIM has been under development since 1990 and was to be started up in January 1994. It will replace the requirements for emission reductions embodied in more than 30 existing and more than 12 future rules. If RECLAIM is successful

---

**BOX 2-8
MANUFACTURERS' BUBBLE**

EPA has a bubble policy that allows existing sources to trade emission credits with a facility. The bubble concept can also be extended to final products such as paints and other coatings. However, regulators have never been enthusiastic about the program. While environmentalists, regulators, and industrialists have expressed substantial support for RECLAIM, the manufacturers' bubble concept has languished.

The goal of the manufacturers' bubble is to limit the aggregate pollution potential of the goods that manufacturers produce and sell in the basin. Since the manufacturers of these products, most of them paints and finishes, would be responsible for annual reductions in emissions, the products would not be accounted for under RECLAIM. In theory, manufacturers would reduce their pollution from a particular coating more than is required in an effort to sell more of a particular product. The incentive would be an increase in sales.

Several problems have slowed or stopped development of this program. First, as envisioned, the program applies only to coatings manufactured and sold in the District and ignores coatings that enter or leave the District. In particular, manufacturers claim they cannot track how products enter the Basin. For example, products may be shipped from other states to California. Overcoming this difficulty has led to the design of pilot programs. This disparity would give outside firms a cost advantage. Second, the regulatory system has no way of accounting for the cans of coatings user firms have in their inventories. Air quality would be affected if they drew down their inventories, although this would not appear in the program's accounting. These problems have been serious enough to curtail the development of the manufacturers' bubble.

---

with NO_x and SO_2, the SCAQMD might extend the concept to VOCs. (See also box 2-8 for another new initiative being considered by the SCAQMD, a manufacturers' bubble.)

How RECLAIM Will Work

The main innovative features of RECLAIM are that it uses marketable permits for emissions, rather than marketable ERCs, and it applies to both new and existing sources of NO_x and SO_x emissions. The principal focus, however, is existing sources.

The name of the marketable permit is the Reclaim Tradable Credit (RTC). Under RECLAIM, the SCAQMD will issue marketable permits to companies for a quantity of emissions. The permits will specify a declining number of emissions over time (that is, 50 tons per year in 1996, 45 tons per year in 1997 and so on). If a company needs to emit more than its allocated RTCs, it may buy marketable permits from another company that has reduced its actual emissions below its allocated level. The SCAQMD believes that RTCs will offer both

---

BOX 2-9
BENEFITS OF RECLAIM RELATIVE TO OTHER APPROACHES

RECLAIM offers four benefits over other approaches to air quality control:

(1) It will enable the Basin to meet or exceed the projected AQMP emission reductions and air quality improvements with a lower or similar impact on jobs, costs, and public health.
(2) Industry will have more flexibility in deciding how to meet its pollution requirements, and facilities will be better able to undertake long-term planning because future pollution control requirements will be better understood and firms will have more control over managing their emissions.
(3) Each facility will be able to choose and implement the most cost-effective strategy to meet its reduction obligations.
(4) For the first time, the emissions from a facility will be capped, a measure that will assure compliance with the annual emission allocations. The required improvement in facility cap monitoring will in turn result in a better understanding of emissions and air quality.

---

new and existing facilities a simple and cheap alternative for meeting some of their pollution control obligations (see box 2-9 on the benefits of the RECLAIM approach).

RECLAIM will apply to stationary sources that hold permits for equipment or processes that emitted more than 4 tons of NO_x and SO_x since 1990. Sources such as equipment rental facilities and essential public services (including landfills and wastewater treatment facilities) will be excluded. Facilities located in areas outside the South Coast Air Basin but still under the SCAQMD's jurisdiction will also be excluded, although facilities in certain industries may enter the program voluntarily. That is, they will be able to use this cost-effective means of reducing emissions only if they opt in. Once a facility joins the program, it will not be able to return to traditional command-and-control regulatory systems. RECLAIM will, as noted, establish a cap on facilities' emissions and specify an annual rate for further reductions. Each facility will get a single permit that encompasses all its emission sources.[8] It will also get an annual allocation specifically for emissions of NO_x or SO_x. The allocations will cover equipment such as boilers, furnaces, and dryers, as well as equipment that is not required to have an operating permit. The allocation will be calculated by determining the historic emissions of NO_x and SO_x by each piece of equipment at the facility and then subtracting the required reduction in emissions. The resulting figure is the permissible level of emissions. To determine historic use, the SCAQMD will evaluate the consumption of inputs by process unit in the peak year of use for each facility. To establish the level of the required reduction, the total reduction required under all rules by December 31, 1993, was calculated for each type of equipment and then aggregated as the required decline. Each facility will be required to reduce its emissions consistent with the specific annual targets. The targets for emissions reduction at each facility are equivalent to what the facility would have achieved under continuation of the command-and-control program and ultimately the AQMP target for 2003. That is, a key feature of RECLAIM is that it commits all the facilities in each market—the market for NO_x or SO_x RTCs—to achieve the emission reduction targets of the AQMP. This should satisfy the SIP requirement to move expeditiously toward attainment.

The RECLAIM annual emission reduction targets will replace the reduction requirements specified in the AQMP and source-specific rules. The rate of reduction for both NO_x and SO_x will follow a straight-line decline from the starting and ending allocations. Facilities can achieve additional reductions in emissions by adding new controls or by modernizing or improving processes. Those companies that find compliance too expensive can purchase lower-cost RTCs achieved by another facility; otherwise, they must achieve the reductions on-site.

The methodology for establishing allocations for facilities was designed to be fair and equitable. With respect to fairness, one goal of the methodology is for the allocation to accommodate necessary operating levels. With respect to equity, the allocations should recognize previous emission control and reduction efforts. Finally, the methodology, besides being fair, equitable and efficient, has to ensure that emission reductions will satisfy the requirements specified in all rules and the AQMP.

Applicable source-specific command-and-control rules will remain in effect for pollutants and equipment not covered by RECLAIM. In addition, sources covered by RECLAIM will still have to meet requirements such as inspections, operations, and maintenance standards. Facilities will be required to maintain existing emission control equipment.

Under RECLAIM, all new or relocated facilities will be subject to the NSR requirements, as well as air quality modelling to determine the concentration of emissions. In addition, they will have to provide RTCs at a one-to-one ratio for the first year. To continue operating, they must have sufficient RTCs to

---

[8] The District divided the universe of firms participating in RECLAIM into two cycles. The Cycle 1 facilities have a compliance year extending from January 1 through December 31; recently the District sent them their facility permits. Ultimately the permit system will be fully automated, which should save considerable staff resources in implementing the program.

mitigate emissions on an ongoing basis. New facilities will not be subject to an annual reduction rate, but they must fully mitigate all emissions by using RTCs to offset their new emissions. Modifications to existing RECLAIM facilities that result in an increase in emissions are also subject to technology requirements and modelling of ambient impacts. In addition, they must demonstrate that the increase in emissions can be mitigated through offsetting internal to the facility or through the acquisition of sufficient RTCs for five years.

All RTC trades conducted to mitigate increases in emissions over the facility's starting allocation and increases in emissions from new or relocated facilities will be subject to the Sensitive Zone provisions of the California Health and Safety Code, which are intended to prevent so-called hot spots. These provisions require that hot spots not be created by the aggregation of new emissions in one spot. It is proposed that the South Coast Air Basin have a Coastal Zone (Zone 1) and an Inland Zone (Zone 2). Zone restrictions will apply only to trades involving a new or relocated facility or a facility that exceeds its starting allocation. A facility in the Coastal Zone may obtain RTCs that originate only in the Coastal Zone. A facility in the Inland Zone may obtain RTCs from either zone.

RTCs will be bought and sold in terms of pounds per year. Although the District will not regulate the market or control prices, it will track the RTCs and prices in an official RTC registry. The District has developed an electronic bulletin board system that will allow RECLAIM participants to know how many RTCs other participants hold. The system can also be used to announce bids and sales.

All companies with a RECLAIM allocation can buy, sell, trade, or otherwise transfer all or portions of their allocation in any given compliance year provided they follow the protocols and reporting requirements. Facilities can purchase credits to meet their emission reduction targets or increase their annual allocation to meet operational needs. Facilities wishing to sell RTCs may do so without prior approval from the District. A seller will be liable for the increase in emissions at the purchasing facility, as well as the decreases in emissions at its own facility, and will be subject to sanctions if it does not have sufficient RTCs for its emissions. Facilities that purchase the credits can either hold them in the form of a certificate, apply them to meet their emission reduction requirements, or apply for an increase in their annual allocation. The seller will register an automatic decrease in the RTCs contained in its facility permit. The buyer's facility permit will automatically be increased when the RTCs are noted for use to meet emission reduction requirements. When the buyer wishes to use RTCs for a new source at the facility or to increase its annual allocation, it must obtain an amendment to the facility permit. Any facility that exceeds its annual allocation without securing the necessary RTCs to mitigate the emissions will be in violation of its annual allocation.

Those permitted to participate in the RECLAIM trading markets are RECLAIM facilities, brokers, non-RECLAIM facility permit-holders, and other individuals interested in trading RTCs. Facilities will be able to trade their emission credits freely at any time, so long as they have registered the transactions with the District.

RECLAIM facilities will also be able to use emission reductions from mobile sources that have been created beyond the requirements of any mobile source rule. The first rule for mobile source reductions is Rule 1610—Old Vehicle Scrapping, which allows for the creation of credits by encouraging owners of old vehicles to retire them earlier than planned. Such early retirements result in unanticipated emission reductions and therefore credits that are available for trading.

Additional mobile source credit programs are planned. For the near future, examples include replacement programs and on-site zero emission programs.

How the Program Was Designed

The development of RECLAIM has taken place in four phases. In the first phase, a public workshop was held in October 1990 to obtain input about the concept. Subsequently, the SCAQMD prepared a concept paper. Phase two began with a second public workshop that was to review the phase one work and provide further input. Following the workshop, the SCAQMD prepared a proposal for a conceptual program. The Governing Board authorized a full-scale feasibility study at its February 1991 meeting. The feasibility study was conducted in phase three. A SCAQMD team evaluated numerous design alternatives for the program and developed five working papers and a summary report with recommendations. The key recommendation was that the District proceed to develop rules for the RECLAIM program. The Governing Board reviewed the material and, after a lengthy public hearing that consisted of six hours of public testimony, directed the staff to proceed with phase four, the development of a series of rules and documents to implement RECLAIM for three program markets—NO_x, SO_x, and VOC. In February 1993, the District decided to separate the VOC market from the hearing schedule for the NO_x and SO_x RECLAIM proposals. The reason was that the enforcement and reporting issues for VOCs are significantly different from those for combustion source emissions and would require additional time and work to resolve. Environmental groups have been concerned about toxic hot spots associated with RECLAIM, and industry has been concerned with proposed reporting. In the meantime, the District will go ahead with command-and-control rules for these VOC sources. Several committees were set up within the SCAQMD, including ones for SO_x, NO_x, and VOC; for penalties/enforcement; and for trading. These committees were responsible for developing administrative procedures. Committee membership cuts across all divisions within the District; notably, representatives from both information systems and stationary source compliance regularly participate in all meetings. During 1993, 40-50 full-time District staff worked on RECLAIM. The number is expected to increase, as staff are shifted from other programs to begin administering the permit system: the FY1992-93 Draft Work Program allocates 55-65 full-time equivalent positions

60   *The Environment Goes to Market*

to RECLAIM. The draft document also lists numerous outreach initiatives, many of which are believed to be related to RECLAIM.

The development of RECLAIM has intentionally been the most open process the District has ever engaged in (see box 2-10). The District hoped to

---

BOX 2-10
PUBLIC PARTICIPATION IN RECLAIM'S DESIGN

There was extensive public participation by virtually every stakeholder interested in the design of RECLAIM. This process was unparalleled in the development of environmental regulation:

- As a part of the feasibility study, the District formed a broad-based advisory committee made up of environmental, industry, political, technical, and legal interests. The committee was to assist in designing RECLAIM. The objectives of the advisory committee were to identify and assist in analyzing key design issues related to state or federal regulatory constraints, to evaluate proposed solutions for a program that would achieve compliance with federal and state clean air standards, and to identify areas of concern. A steering committee was formed out of the advisory committee to assist District staff on key issues in the proposal's design. However, both committees played integral roles in the development of the program.

- The District followed a cooperative approach to rules development. The steering and advisory committees were expanded to include additional representatives when the rules were being developed and continued to meet during rules development to provide input into program requirements. Industry and environmental proposals were given serious consideration and contributed to the design of RECLAIM. For example, the approach to allocation, which embodies the concept of relying on past throughput levels, was developed by one of the key industry groups, the regulatory flexibility group. At the beginning of rule development, several working groups were formed to provide input on various elements of the program. By all accounts, the rule-making was as open as it could be. The acceptance and use of this cooperative approach were critical to gaining industry's endorsement of the program.

- A similarly cooperative approach was followed in developing the emissions estimation techniques, or protocols, which describe how emissions will be monitored and reported. There was extensive interaction with industry and environmental groups. The policy choices made and final form were critical to gaining EPA's acceptance of the program. The protocols provided EPA with the opportunity to endorse the program and make some CAA interpretations since the protocols require significant advances in measuring techniques to allow the program to be monitored and tracked.

- Each week, the SCAQMD held public meetings in which the committees of District staff and representatives of stakeholder groups discussed various aspects of the emerging program. The stakeholders included large businesses, small businesses, environmental groups, EPA, and the ARB.

achieve two goals through extensive public participation, particularly the weekly public committee meetings. First, by gathering as much information and as many opinions as possible while the program was under development, it wanted to minimize the number of problems that it would have to deal with later. Second, it hoped to get contending stakeholders to buy into the program. The successful adoption of the RECLAIM program proved this approach to be correct.

Information Systems

RECLAIM requires a number of different computerized information systems, and information resource specialists have been actively involved in the planning and design phases. In the case of NO_x, a continuous emissions monitoring system (CEMS) has been established. Remote terminal units (RTUs) will monitor emissions at the source and send hourly reports directly to the District's central data processing unit (CPU). The CPU runs continuous checks for anomalies, comparing current emissions with those in previous periods, and also compares the emissions with those from similar plants in the Basin. Anomalies are immediately brought to the attention of enforcement personnel. In the cases of SO_x and VOC, the District sends mailers to permit-holders requesting emissions data, which they return for entry into the system.

The RECLAIM information systems have to meet three criteria: reasonable development costs; sufficient sophistication to provide the required information; and ability to validate compliance. As to the first, the costs of developing the systems are declining as the District becomes more experienced with design and implementation of management information systems. The systems need to be sophisticated because they will be used to analyze and use data concurrently. Concerning compliance, the computer system will, as noted, actively search for anomalies relative to a firm's historical trends or relative to trends at similar firms in the District. If it finds an anomaly, it alerts enforcement personnel and generates data in the short run. These data allow a quick response to the potential violation and generate reports that are robust enough to permit successful prosecution of facilities that fail to comply.

There is little difference between the information system required for command-and-control environmental regulatory systems and that for a market-based environmental regulatory system. In both cases, the goal is to monitor compliance in a streamlined manner while minimizing reporting requirements. The systems under development for RECLAIM have been driven by the monetization of the benefits of pollution control and the creation of a quasi-property right.

RECLAIM Versus NSR: Similarities and Differences

This section compares the SCAQMD's traditional NSR regulatory approach and the impending RECLAIM program. In making this comparison, it must be borne in mind that as of this writing RECLAIM had been adopted (on October 15, 1993) but had not yet been implemented. Final details on its design and operations were not available, given that even adopted rules can be modified.

Table 2-5 summarizes some of the anticipated similarities and differences between RECLAIM and the NSR marketable permit trading under offset requirements (the entries in bold are discussed below). RECLAIM departs from previous command-and-control practices in three major ways. First, RECLAIM extends and standardizes the bubble concept. Historically, the District did not support EPA's bubble policy. Industry also was confused about supporting this policy. While the bubble concept gives industry considerable flexibility to meet emission limits, it requires firms to cap emissions and assume specific

TABLE 2-5
COMPARISON OF RECLAIM AND NEW SOURCE REVIEW (NSR)

ISSUE	RECLAIM	NEW SOURCE REVIEW
Coverage	390 firms	Very small
Unit of regulation	**Industry site (bubble)**	**Permitted equipment**
ERC trading purpose	Meet annual reduction goals and NSR requirements	Expand or establish new sources
Economic impacts	Lowers cost to industry	Minimizes siting cost
Health impacts	None	
Employment impacts	None	
Rule development style	Participative/cooperative	Hierarchical/adversarial
Rule development procedure	**Once and for all**	**In place under Regulation XIII; frequent revisions**
Monitoring	Continuous passive	Same as for existing sources
Data requirement	Substantial	Slight
Media outreach function	Build credibility for market/defend penalties	Minor
Source of reductions	**Annual reductions in total emissions**	**New expanding source**
Basis for District revenue generation	Fees according to each firm's proportion of all emissions	Fees according to each firm's accrual emissions

Note: Entries in boldface are discussed in the text.
Source: National Academy of Public Administration

permit limits. Because the RECLAIM program and operating permits mandated under the 1990 CAA impose emissions caps for facilities, bubbling has become a viable compliance strategy for NO_x and VOC emitters.

Second, in designing RECLAIM, the District has abandoned the traditional rule-making procedure for existing sources, a drawn-out, resource-intensive, and contentious process that often involves lawsuits for even minor changes in regulations. Instead, the intent is that RECLAIM will require few mid-course corrections until 2010. Although initially the development of the regulations under RECLAIM was more contentious than regular rule-making, once the system is in place, it will not need frequent tinkering. The reason is that under command-and-control, firms are regulated on a source-by-source basis one at a time, whereas RECLAIM regulates whole classes of firms—NO_x or SO_x, or VOC emitters—at one time. District staff hope the RECLAIM approach will preclude the need for further major efforts of this type.

The third major difference is the across-the-board annual cuts in emissions required of all participants. Historically, facilities achieved many emissions reductions primarily through shutdowns and the voluntary creation of ERCs. Because attainment of federal and state requirements means an 80 percent reduction in stationary source emissions, RECLAIM is instituting this challenging (some would say impossible) annual reduction. Now many reductions must come from retrofits unless a substantial part of the inventory is shut down.

There are other differences as well. For example, with respect to generating revenue, instead of charging firms for their actual emissions, RECLAIM will charge firms for the proportion of each type of air emission they control. In that way, as emissions decrease, the District will not see its operating funds diminish as well.

Challenges in the Implementation of RECLAIM

Now that RECLAIM has been approved, the District will have to address a number of challenges as it implements the program. The devil is in the details, which do not capture either media attention or stakeholder involvement. As yet, the administrative and oversight components associated with RECLAIM are not fully understood.

Revenues and Expenditures

Thus far, the SCAQMD has performed only external cost analyses on RECLAIM. That is, it has estimated just the costs to businesses and has not undertaken internal cost projections. The internal finances, however, pose many questions. For example, will RECLAIM consume a disproportionate share of the SCAQMD's budget? In the latter half of the 1980s, the District was able to rely on major increases in emissions fees to fuel its growth. This source of revenue is expected to stabilize in the 1990s. While all the evidence is that

RECLAIM is the most cost-effective way to control emissions, it is also hard to tell if RECLAIM will in fact be excessively expensive or very cheap as a means of improving air quality. The cost to regulators is also an open question. Cost estimates must be carried out at the level of the divisions as well. For example, will there be a greater or lesser demand for enforcement resources for the chief prosecutor's office? Should an internal cost analysis have been an explicit part of the committee-based development work? Regardless, should internal costs be tracked as part of an ongoing evaluation plan and, if so, how?

Staffing

A market-based system such as RECLAIM changes the focus of many activities. First, since many source-by-source regulations are not going to be developed, those engineers with expertise in this area will not be needed for rule-making. Where will these people go within the SCAQMD? Some staff previously involved in RECLAIM are being reassigned for rule-making on non-RECLAIM sources, of which there are many. However, it is clear that as case-by-case rule-making declines, the need for rule-writing specialists declines. Second, as data collection, organization, and analysis become more important in tracking RECLAIM credit trades, more emphasis must be placed on management information system-related activities. Command-and-control approaches put a premium on certain types of human capital—educated and experienced engineers. For a marketable permit system to work well, the regulator must shift the focus to data and informational activities and perhaps elevate compliance management to a priority. What happens to displaced SCAQMD engineers? Will those threatened with a job loss fight a rearguard effort to modify RECLAIM? How will this change affect the SCAQMD's organizational culture and morale?

As noted, there is a perception that performance reviews have not always been taken seriously. The District has been criticized over the years for its inattention to its relations with industry, and the importance of good regulator-industry relationships has hardly been mentioned in connection with RECLAIM. A market-based system will require improved relations between the SCAQMD and businesses. One question is whether the performance review process needs to be revised to reflect the principle of customer focus. Should reviews of employees' performance take into account responsiveness to businesses and the development of close working relationships?

District staff interviewed during this study believe there will always be another innovative project to work on once RECLAIM is routinized. That belief raises the question of who will manage RECLAIM day by day. Even though the SCAQMD has been restructured with the goal of decentralizing authority, if those who developed RECLAIM—the people with the deepest understanding of the program—move on, its management, which will consist largely of new personnel, may be less able to resolve issues. Decision-making may be stalled as issues move up through the ranks of management in search of resolution.

Organizational Culture

While the District's innovative culture has helped it accomplish many important goals in the area of command-and-control regulations, businesses prefer *market* structures characterized by certainty and stability. As envisioned, under RECLAIM the regulatory role of the District with respect to the 400 firms participating in the NO_x trading program and the 40 participating in the SO_x trading program may involve little more than monitoring compliance, monitoring the NO_x and SO_x markets, and responding to anomalies. This type of work requires consistency and rapid response, not necessarily innovation. In short, the culture needed to develop RECLAIM may not be the best one for its execution.

While there are some things about the SCAQMD that are not likely to change, such as the location of its headquarters and sources of revenue, others might change. Tracking the impacts RECLAIM has on the District is an important component of a potential evaluation system, so that managers can identify unintended impacts and deal with them. Some people have suggested altering the tone and tenor of the governing board's meetings, with attention to cooperative solutions to obviously complex problems. Perhaps moving away from contentious source-by-source writing of regulations will moderate relations between industry and regulators and lead to the establishment of a more customer-focused administration.

Industry Support

The cost savings for industry associated with RECLAIM arise from the general application of the bubble concept and the opportunity to trade emission reductions. In any market, predictability and consistency of regulation are critical to industrial participation. Here, the political nature of the regulatory process makes the District's effort to behave consistently and predictably all the more important.

With respect to the committee-based development of RECLAIM rules, while it appears the goal of information-gathering has been achieved, the goal of buy-in has not, although the SCAQMD will never achieve total buy-in. Interest groups, rather than buying into the process, appear to be defining critical issues; staking out turf; and girding for future political, regulatory, and legal battles that will ultimately define key aspects of the program. Conversely, some non-District professionals involved in the design of RECLAIM are concerned the District is not including diverse groups in decision-making. There is concern that true decision-making takes place internally or between one or two industry representatives and District staff and that the District uses the public meetings to legitimize that function.

Concerning reporting, the District currently relies upon volumetric data, often measured in parts per million (ppm). These data can be generated in a short period. RECLAIM, on the other hand, requires reporting of mass concentrations such as tons per year. Mass concentrations are a function of concentration and throughput. Generating the samples upon which calculations are based takes a lot of time. Auditing a source's manual record-keeping may prove costly and difficult for the District.

Boxes 2-11, 2-12, and 2-13, provide three perspectives from different stakeholder organizations. These perspectives illustrate some of the problems and perceptions of industry and give a flavor for how those affected view RECLAIM.

BOX 2-11
REGULATORY FLEXIBILITY GROUP

"The companies in this basin will find a way to do it,
if you give them the reins."

*From the testimony of Robert A. Wyman
(on behalf of a coalition of major businesses), as cited in the*
Environmental Digest *(Fall 1993)*.

We know th[e] command-and-control alternative well. We've seen it drive work from the basin when companies cannot meet technology-forcing rules. We've seen companies large and small spend millions to reduce mere pounds of emissions when cheaper alternatives were available elsewhere at their facilities. . . . Under our current program, we cannot have clean air AND a healthy economy!

But is command-and-control the only way to get clean air? No, emphatically no, it is not. Under RECLAIM, we reward innovation. If a technology reduces emissions, it's worth money. Under RECLAIM, each company will have the same reduction requirements as under the AQMP. The difference is that we can choose which equipment to control and when. I have yet to meet the company, large or small, that cannot save money when it can make the choices. Under RECLAIM, even the smallest business benefits, because it has the choice of whether to install the same controls that would be required under AQMP, to control elsewhere or to buy credits.

RECLAIM does not make pollution control cheap, just a lot less expensive. It doesn't make responsibility for the environment go away; it just gives us the opportunity to exercise that responsibility wisely. . . .

. . . I am here to give you our group's strongest support for RECLAIM. It is a world-class proposal. It will make Southern California the world center for clean air innovation. Other areas of the world will follow suit. . . .

. . . We know how markets work. They have brought us a quality of life unparalleled in the world. But they have a weakness. Until today, they have not reflected the price of economic activity on the environment. That has been a hidden price and, as a result, our industry and our mutual love of the environment, for which each of us is responsible, have been at odds. That need not be so. If you adopt RECLAIM, you will be showing the world how the economy and the environment CAN work together. . . .

> **BOX 2-12**
> **SOUTHERN CALIFORNIA GAS COMPANY**
>
> "It neither makes sense [n]or is in Southern California's best interest to adopt RECLAIM . . ."
>
> **From the testimony of Richard D. Farman, CEO, Southern California Gas Company, as cited in the Environmental Digest (Fall 1993).**
>
> You will be hearing from many of the larger businesses in the next couple of days, most of whom are customers of the [Southern California] Gas Company, expressing their support for RECLAIM. But, I am here to represent the majority of our customers who are opposed to RECLAIM—the small or moderately sized companies. That's because during the past few months more and more of these customers have told us that they do not want to be regulated under RECLAIM. They told us how RECLAIM's complex and, in some cases, unworkable provisions could stop their expansion and restrict their recovery from the recession. They told us how they would have to add staff or hire consultants and lawyers to manage the complicated and unnecessary recordkeeping requirements. Expenses such as these do not help the economy, and do not help clean the air . . .
>
> . . . [A]sk yourself whether RECLAIM is fair. Can your Board adopt a program that would give some industries a 90% reduction in compliance costs while, by the District's own numbers, it would increase costs for other industries by up to 600%? . . .
>
> We offer the proponents of RECLAIM an alternative: If RECLAIM is as desirable as they claim, then make the program voluntary! Let those who want RECLAIM, live under its requirements, while others can choose to stay under command and control regulation of the 1991 AQMD. I'm not suggesting that anyone be let off the hook, rather than we each choose our own medicine.

Possible Litigation

During the RECLAIM hearings, staff of the California Air Resources Board (CARB) said they found that the program met applicable state requirements, as specified in AB 1054, the key state law governing the program. Signed in 1992, the law is directed toward market-based control programs. It sets forth specific criteria that such programs have to meet, one of which is that the program must result in emissions reductions equivalent to those that would have occurred had the AQMP continued to be implemented.

Representatives of Citizens for a Better Environment have said they will sue the CARB on AB 1054 grounds. Other groups such as the Natural Resources Defense Council (NRDC) have said they will closely scrutinize EPA's findings on the program. In short, everyone is tracking the program. The first report back to the District's Governing Board is scheduled for April 1994.

> **BOX 2-13**
> **SOUTHERN CALIFORNIA EDISON**
>
> "We believe that the AQMD has created a new model.
> Its name is RECLAIM."
>
> *From the testimony of Bryant Danner, Senior Vice President, Southern California Edison, as cited in the* **Environmental Digest** *(Fall 1993).*
>
> Southern California must continue to clean up its unhealthful air. Edison believes that RECLAIM is by far the best way to achieve that goal without threatening the future economic prosperity of the region. The program integrates the need for clean air and a healthy economy.
>
> We urge the governing board of the AQMD to approve RECLAIM for these reasons:
>
> - First, the program will clean up the air at reduced costs. The approximately 400 businesses covered in RECLAIM will save $93 million a year on air pollution control costs compared to what they would spend under command and control regulations.
>
> - Second, RECLAIM will save about 1,700 jobs in the LA Basin between 1994 and 1999, compared to command and control regulations.
>
> - Third, it encourages Southland companies to develop new pollution control technologies that can be marketed around the world, stimulating the Southland's economy.
>
> Finally, RECLAIM allows all participants the flexibility to develop strategies that will work best for them, depending on their individual circumstances and business needs. . . .
>
> We are serious about our responsibility to help clean the air. We are equally serious about helping to improve the economy of the region for the benefit of our customers. We support RECLAIM because we believe it's the path to simultaneously achieving both of these goals.
>
> The cost savings Edison realizes from RECLAIM will be passed on to our 4 million customers, further helping the region's economy. Many of those 4 million customers are small business owners.
>
> Small businesses in the program also have the flexibility to either buy credits or choose their own methods of reducing emissions.

Evaluation

It appears the SCAQMD has not yet developed an evaluation plan for RECLAIM. That challenging task will have to be undertaken and brought to bear on the operations of the program. At a minimum, an evaluation plan should include the questions to be asked, procedures for conducting the evaluation, and an estimate of costs.

Some of the questions that are likely to be included in the preparation of an evaluation plan for RECLAIM are the following:

- Are the goals of RECLAIM realistic?
- How is the appropriateness of RECLAIM's goals determined?
- Is RECLAIM easy to use as measured by transaction costs, including time required to perform key activities?
- Is RECLAIM cost-effective as measured by the costs of RECLAIM credits and cost comparisons of RECLAIM credits versus credits of RACT, BACT, and LAER?
- Are fewer FTE administrative positions needed for RECLAIM than for other programs?
- What are the key activities regulators will spend their time working on in conjunction with RECLAIM?
- Has RECLAIM reduced the burden on industry in terms of administrative costs, permit delays, and uncertainty?
- Has RECLAIM promoted new ways to control emissions?
- Has RECLAIM resulted in less deterioration in air quality as measured by hot spots, indicators of emissions, and information about ambient air quality?
- Has RECLAIM resulted in more-expeditious progress toward the achievement of air quality objectives in comparison with traditional approaches?
- Is RECLAIM less burdensome administratively than other programs with respect to enforcement actions and management information systems?
- What are the intended and unintended impacts of RECLAIM on the District's organizational structure and behavior?

This list is not exhaustive. The questions do, however, approach RECLAIM from three contrasting perspectives: a goal-oriented one, a process-oriented one, and a results-oriented one. For example, some of the questions focus on objectives; others on resource and information flows, managerial discussions, and inputs; and still others on outcomes and performance levels. All three sets of questions incorporate comparisons with other regulatory programs. All three, it should be noted, are at the heart of managing the program.

These questions may also serve as the basis for:

- Measuring degrees of progress toward levels of attainment.
- Demonstrating the strong and weak points of the program.

- Examining the adequacy of RECLAIM in comparison with SCAQMD's total needs.
- Providing quality control.
- Generating new and improved processes and procedures.
- Indicating the transferability of RECLAIM to other areas.
- Advancing the evaluation skills of the staff.
- Providing public accountability.
- Developing a critical attitude among program staff.

The details of an evaluation plan for RECLAIM will not be spelled out here because to do so would go beyond the data at hand. However, it should be pointed out that an evaluation of RECLAIM cannot be performed without developing an understanding of how RECLAIM fits into other SCAQMD programs. There are three ways to look at this issue. First is how RECLAIM complements or conflicts with other programs. For example, does it conflict in terms of timing? Does it take resources away from other programs? The second aspect emphasizes maintenance of the balance between the functional responsibilities of the District. The third takes into account how RECLAIM and other District programs interact over time in a cumulative way.

Other Issues

At both EPA and the ARB, regulators have said they are willing to reinterpret and/or amend the statutes to allow RECLAIM to move forward. One regulatory hurdle involves reinterpreting the requirement for RACT for all emitters in a non-attainment area. RACT is the reasonably available pollution control technology or pollution control level for existing sources in non-attainment areas. Historically, RACT was determined on an emissions point basis. Under RECLAIM, RACT goals would be considered in the aggregate. Other rules will have to be reinterpreted/amended to allow for mass- rather than volume-reporting to the ARB and EPA. Finally, rules that may hinder the cost-saving elements of RECLAIM, such as the District's ability to sequester ERCs to demonstrate reasonable further progress, must be changed. In sum, state and national regulatory requirements could delay implementation or alter the structure of RECLAIM.

Staff members, as noted, describe RECLAIM's development as a once-and-for-all regulatory effort. They anticipate only rare mid-course adjustments to ensure the air quality goals set for 2010 are met. Outside parties take a different view. They believe the local economy will be hurt by the effort under RECLAIM to achieve an 80 percent reduction in stationary source emissions. As the local economy is affected, there will be pressure to amend the program. This pressure will require a political response, and RECLAIM may not be preserved in its current state.

Finally, given its powers and mandate, the SCAQMD as regulatory agency has a high media profile. When industry and the environmental community are unhappy, this high profile can be a burden. Not only does the District actively put out its own messages to the media, but it also monitors and reacts to media coverage of its activities. In the case of the NSR, the issues regarding rulemaking have often been contentious. The editorial board of the Orange County *Register*, a conservative paper in the basin, has often criticized what it sees as undue intervention by the SCAQMD in industrial activities. While it is too early to tell, the media issues that RECLAIM will have to address may involve building support for the emissions trading market or justifying the penalties that accompany violation of RECLAIM rules. Now that RECLAIM has been approved and trading will soon start, the media are likely to become the vehicle for the translation of RECLAIM for the public.

CONCLUSIONS

This case study highlights the difficulty of redirecting an environmental regulatory program onto a different course, in this case from a command-and-control approach to a novel, broad-ranging market-based emissions trading program. It also describes some of the administrative aspects of the program and points out the need for its continual evaluation. Among the lessons provided by SCAQMD's experience are the following:

- The sponsoring agency must be able to project significant improvements from the new program through a series of meetings with numerous stakeholder groups. This was achieved by the SCAQMD through three years of face-to-face meetings with all stakeholder groups and the dissemination of the results of contract research highlighting the economic outcomes.

- When switching from a command-and-control view of the world to an economic incentive one, the District must come up with a design that satisfies external stakeholders, such as small and large businesses; environmental groups; ethnic groups; and local, state, and federal regulators. Even if the existing system does not work well, doing something else that produces the same degree of dissatisfaction will not be acceptable.

- The sponsoring agency must integrate the new program into its organizational structure, budget, information systems, career paths and evaluations, staff allocation, training, industrial relations, and culture.

- The new program must not only meet a societal cost-effectiveness test; it must meet an internal cost-effectiveness one as well. That is, can the agency afford the program?

The SCAQMD is currently grappling with many issues and will continue to do so as it moves from the design phase to the implementation phase. It will complement its traditional economic incentive-based NSR regulatory program with a system that takes more advantage of industrial initiative and market forces. Such programs have been shown to provide useful supplements to traditional regulatory systems. Although the RECLAIM program has benefitted from consultation with external constituencies that have been supportive, as well as from its tradition of innovation, organizational flexibility and strengths in such staff areas as information systems and ecological-economic modelling, the program may be negatively affected by the District's strong reliance on its own expertise, its rejection of externally generated information, and its strong desire to design a program that is not susceptible to change. These are the ingredients that will determine whether RECLAIM is a regulatory success or failure.

A challenge the District faces in implementing RECLAIM is modifying its culture, which traditionally has not been amendable to input from outside constituencies. In developing RECLAIM, the District has had to get considerable input on a range of issues: baseline data, rates of reduction, trading of commodities, applicable sources, and so on.

Moving into implementation will require the same cooperation across regulatory-environmental-industrial constituencies that marked the development of RECLAIM. Without the extraordinary degree of cooperation exhibited by all stakeholders, RECLAIM would still be a twinkle in the eyes of regulatory reformers. The successful implementation of RECLAIM, especially the development and implementation of evaluation systems, must be predicated on a similar level of cooperation. To summarize, the SCAQMD has demonstrated what regulatory reformers have always known: an aggressive and well-trained regulatory agency can effect change if the reform is sound, the process is inclusive, and the agency is willing to put in the substantial up-front sweat equity needed to make the program happen.

CASE STUDY:
THE RUSSIAN POLLUTION CHARGE SYSTEM

In 1991, the Russian Federation initiated a pollution charge system to curtail pollution. Implementation of the system followed a series of pilot programs that had begun in 1988 under the former Soviet Union. The pollution charge system marks the first time Russia has used economic incentives or a market-based approach to promote environmental policy.

The system involves charges on the discharge of pollutants into the air, water, and soil. The government collects the charges as taxes and deposits them into special, non-budgetary "environment" funds at the central, regional, and local levels; they are then used to finance environmental cleanup and abatement technologies. In addition, polluting enterprises are required to invest in and operate pollution control and treatment programs. Finally, the federal, regional, and local governments allocate a share of their budgets to environmental protection and conservation programs supporting environmental and natural resource objectives.

The Russian system of pollution charges is remarkable in several ways:

(1) First, the charge base is large, covering hundreds of pollutants.

(2) Second, only in Russia; and perhaps in Poland, are expenditures for environmental protection so dependent on regional environmental funds.

(3) Third, and perhaps most surprising, on the one hand the government has issued detailed instructions on how to collect pollution fees from firms and other economic agents. On the other, it has not developed the legal and institutional framework necessary to carry out this system. There is, for example, a shortage of trained regulatory personnel, uncertainty over property rights and economic liability regarding the use of natural resources, and a lack of generally accepted measurement standards and equipment to support monitoring and enforcement.

(4) Also remarkable is that Russia is adopting the pollution charge system during the painful transition to a market economy. Inevitably, the system combines elements of the socialist past with the newest incentive-based approaches to environmental management. Market-based approaches are new even in developed countries. Although the United States and some European nations have levied emissions charges, most apply to water or hazardous waste streams. None is as ambitious as the Russian system.

This section describes the basic features of the Russian pollution charge system. It opens with a description of how the system came to be and how it is structured. The environmental and economic situation in Russia; the basic principles of the charge system; the organizational structure within which the system fits; and the main features of the system, including details on the calculation of the charges, the collection and allocation of revenues, and program management, are all described.

The environmental funds are addressed next in terms of the spending of revenues and management of the funds. That discussion is followed by a review of selected administrative aspects of the program: monitoring; auditing and enforcement; personnel training; oversight and evaluation; the costs of implementation; and attitudes toward the program.

This section concludes with a brief look at some program results and offers recommendations on evaluation of the pollution charge system.

Several points emerged from this study:

- The system suffers from the following problems: a weak infrastructure for handling the complex planning, analysis, monitoring and enforcement, litigation, and inter-jurisdictional negotiations involved in implementation; a shortage of trained personnel; and a lack of comprehensive information on the hazards posed by specific pollutants at individual plants.

- The Law on Environmental Protection, which governs the levying and collection of fees, needs strengthening. For example, the law provides local governments with considerable enforcement flexibility. If a firm is persuasive in claiming that a pollution charge will cause it to close, the local government can reduce or waive the fee. Moscow provides no standards for making these decisions. A further problem is that companies sometimes fail to pay the emission charges, and foot-dragging can be as successful a way to avoid environmental compliance in Russia as it sometimes has been in the United States.

- There is a lack of financial and legal regulations governing the exploitation of resources. Command-and-control types of regulation require a solid regulatory infrastructure. In the former USSR, however, permits, impact statements, monitoring protocols, chain-of-custody requirements, bonding, laboratory licensing, insurance requirements, and other legal, financial, and quality control tools were either unknown or had been only newly implemented. Similarly, economic incentive-based programs must be grounded in a solid infrastructure, especially for monitoring and enforcement, and suffer from its absence.

- The funds generated by the pollution charge program are not adequate to remediate existing environmental problems.

- Control of numerous pollutants is to be achieved through fees. The overall system does not adequately use other options such as technological and product substitution, as they are not feasible for financial reasons and because of an absence of regulatory drivers.

- No oversight or evaluation mechanism exists to provide environmental managers and other stakeholders with real-time information about the status of the pollution charge system.

- Russian observers judge many facets of the experiment to be successful, although their measures of success are largely subjective.

HISTORY OF THE RUSSIAN POLLUTION CHARGE SYSTEM

The pollution charge program is the first experiment with economic incentives for environmental management in Russia. The history of its development, basic features, and results to date are examined below.

Background

Acknowledgment of the role of environmental factors in social and economic development and the need to account for the environment began in the former USSR almost at the same time as in the West. Authorities set up environmental health standards that were based on stringent medical and biological criteria. Soviet authorities were proud that their environmental standards were so strict, and they did not pay attention to the fact that nothing had been done to meet those standards. The standards continued to be improved while the real environmental situation became worse and worse. In many cases, air pollution exceeded the standards by a factor of 10 or 100—and water pollution by a factor of 1,000.

On October 27, 1960, the Russian Federation of the USSR enacted its first Law on Natural Resource Conservation. In the second half of the sixties, there was a national debate over the proposed construction of a giant paper mill on Lake Baikal that proved an important step in the growth of the environmental ("green") movement. Although opponents of the paper mill lost, the debate focused public attention on how economic development activities can damage the environment.

The first debates on environmental problems occurred within the academic and engineering circles. However, in 1971, the 24th Party Congress passed a resolution that officially acknowledged the environmental issue, and in 1974 and 1975, the government included special chapters on environmental protection and natural resource use, including emission indices and expenditures for pollution control or treatment, in its economic plans and statistical reports. In the late 1970s and early 1980s, the USSR parliament passed a law to protect the air, along with other acts governing other parts of the environment; in 1978 and 1982, the USSR government, through the Council of Ministers and the Central Committee of the Communist Party, issued two special decrees on environmental protection and the use of natural resources.

To understand Russian environmental and economic policies, it is essential to know that in the former Soviet Union laws did not play the same role they do in the United States and European nations. They did not matter, because of a lack of institutional support and enforcement. They were as nominal as health standards. The primary official documents that affected environmental management in Russia were the "resolutions" of the Communist Party Congresses and the decrees by the USSR government. However, even those decrees did not matter very much. As with environmental laws and ambient standards, they only decorated the surface of the Soviet economic system. The fact was that emission reductions and control technologies were never included in the plan targets.

The laws and decrees set forth a number of tools for environmental management, including penalties for environmental violations. At first, the penalties did not affect the behavior of enterprises because they were owned by the government. To deal with evasion, however, the government decided to fine the directors of the state-owned enterprises personally or to reduce or cancel their bonuses when they sanctioned environmental violations.

The seemingly tough legal measures did not improve the situation. One reason is that stiff penalties were rarely applied. For example, the head of a large state enterprise might be fined only 0.5 percent of annual salary. Sometimes a manager would lose an annual bonus if there were too many violations. However, managers often faced a dilemma—achieving the pollution reductions would mean not achieving production targets. Consistent with the above-mentioned secret and powerful "rules of the game," experienced managers would choose to be fined for excessive pollution in order to meet key production targets.

Before 1987, the government officially was using command-and-control methods to accomplish environmental management. Beginning in the early 1970s, however, the Central Economic and Mathematical Institute (CEMI) of the USSR Academy of Sciences, which was familiar with Western economic concepts, researched the use of economic instruments for environmental protection. The group of researchers at CEMI introduced the concept of a pollution fee as early as 1971 at a conference on optimal planning and management of the national economy. It proposed calculating the charges for emissions on the basis of the value of the marginal costs of reducing the pollution. Those values were not proposed to be used as a real tax to collect monies from polluting enterprises but as a planning tool. In the mid-1970s, Konstanin Gofman, head of the CEMI environmental group, proposed using long-term emission charges as working instruments and allocating the collected funds among the special regional financial organizations. Gofman also proposed combining some of the charges collected with other funding to support pollution control activities. In other words, the principal features of Russia's current pollution charge system first came to light two decades ago in the academic circle of "Western-oriented" economists.

Under the supervision of another economist, Oleg Balatzkiy, another group of environmental specialists in the city of Sumy, Ukraine, developed the concept of economic damages from pollution and elaborated damage-based methods to evaluate environmental externalities (mainly as applied to air pollution). Because Gofman's proposal to charge for pollution did not find any official support at that time, his group also started calculating the costs of environmental damage in conjunction with the cost-benefit analysis of economic development plans and projects.

The work of both these groups greatly influenced the development of the pollution charge program. For more than 15 years, economic evaluation of environmental externalities was the focus of Soviet environmental economic studies and was discussed in the press, at workshops, and in meetings with government representatives. Gradually the notion grew among officials that the external costs of pollution needed to be internalized, at least in feasibility studies. In 1983, the authorized Soviet agencies approved an official methodology for calculating the cost of environmental damages and, once that figure was known, for determining the economic efficiency of investments in pollution control technologies. The methodology was published in the open press in 1986. It did not, however, authorize actual use of charges—their application was restricted to feasibility studies and plans. Nevertheless, several regions of the former Soviet Union chose to use the methods, or modified ones along with other techniques, to determine pollution fees, which they first introduced in 1988.

Until the late 1980s, decision-makers did not intend to charge enterprises for the cost of their environmental damage. However, as a result of the discussion and gradual development of a damage-based approach to polluter liability, by the time *perestroika* occurred, decision-makers had become "morally and instrumentally" ready to accept pollution fees as a new, economic-based instrument for environmental management.

Early Experiments with Pollution Charges

Perestroika, which started in 1985-86, promoted public interest in environmental problems. For the first time, the Soviet public was informed about the scale of the environmental crisis in the country. Official interest in economic regulatory instruments was regarded as a "symbol" of social and economic transformation. In January 1988, the government issued a decree that called for a radical reorganization of the management of natural resources and the environment. It specified the need for pollution charges, including taxes on emissions *within* acceptable limits, or quotas (in addition to fines for emissions that exceeded the quotas, some of which were already in place), and it established a network of environmental funds to receive and allocate the revenues from the charges. The decree was a direct precursor of the pollution charge system the Russian Federation introduced in 1991.

On the legislative side, the first mention of pollution charges occurred in the Soviet Law on Governmental Enterprise in 1988 and subsequently in the Law on Taxes on Enterprises and Organizations in 1990. The latter law established that pollution charges would be paid out of profits (rather than being treated as a cost of production) and set a limit on the total amount of taxes that could be collected. These two laws did not establish how the amount of the charge was to be determined or how it would be collected.

The 1988 decree also established a USSR Committee on Environmental Protection and Natural Resource Use (Goskompriroda), which replaced the USSR Council of the Ministers' Commission on Environmental Protection, which was the former focus of national environmental planning. Goskompriroda had the same authority as a ministry. (Goskompriroda was renamed the Ministry of Environmental Protection and Natural Resources [MEP&NR].) Goskompriroda had a relatively large economic division, which was responsible for preparing regulations governing the operations of the pollution charges and non-budgetary environmental funds. At the end of 1989, it submitted regulations to implement a pollution charge system to the USSR Council of Ministers.

At the beginning of 1990, the Commission on Economic Reform of the Council of Ministers approved the introduction of pollution charges in 49 regions on an experimental basis. The main objectives of the experiment were to discover the organizational and, to a lesser extent, financial possibilities of pollution fees, and to estimate the effectiveness of the charges as an economic incentive and a way to earmark funds for environmental protection. The regions selected were mainly in Russia and Ukraine, where local agencies of Goskompriroda had a relatively strong influence and qualified personnel. The list included the Moscow and Leningrad (now St. Petersburg) *oblasts* and a number of *oblasts* in the areas of the Ural Mountains and Volga River.[9]

Goskompriroda and the Ministry of Finance, which was responsible for tax policy, were in charge of implementing the experimental program. Early on there was a rule, abandoned after one year, that no more than 7 percent of an enterprise's profits could be taxed for environmental protection. The growing trend toward regional independence led many regions to introduce and enforce their own pollution charge systems, which sometimes deviated from the guidance provided by the central environmental authorities. In about half the regions involved, local experts and environmental regulators decided how to impose and allocate the charges. Some regions—Kemerovo, Donetzk, Dnepropetrovsk and Kostroma—set up their experiments in 1988. Others followed in 1989. Several tried to use the 1983 instruction as a guide for estimating the economic damage from air or water pollution and for calculating the charges for emissions or effluents. Regions that had substantial experience in applying the 1983 instruction attempted to introduce pollution charges on their own, in accordance with the 1988 decree.

Regional Experiments

The government eventually tested using the charges to complement command-and-control and administrative approaches to pollution control in a series of experiments at the *krai*,[10] *oblast,* and city levels. The goals were to:

- Refine the methodology for establishing and levying the charges.
- Stimulate industry to take pollution control measures.
- Specify charges specific to effluents or emissions and the damage they caused.
- Differentiate between the charges for the same pollutant on the basis of the environmental differences between regions.
- Establish consistency between the then-current Soviet and Russian laws while keeping them as simple as possible.

A working group of specialists from the Russian Federation was appointed in 1991 to help achieve these goals. Its role was to summarize the experiments and their outcome and, based on its members' experience, to recommend a more rigorous fee system for Russia as a whole. The group included members of the Nature Protection Committees of the *oblasts* of Volgograd, Leningrad, Kemerovo,

---

[9] An *oblast* is analogous to a province with executive and legislative functions.
[10] The same as an oblast with a provision for ethnic minority participation.

Kostroma, and Saratov; the Krasnoyarsk *krai*; and the city of Moscow. The experiments were based on earlier work in Ulyanovsk, Zaporozhye, and Kemerovo involving an assessment of the damage from water pollution, as spelled out in the report "On the Introduction of Temporary Fee Norms for Environmental Pollution and on the Procedure for Forming and Using the Reserve Nature Protection Fund."[11]

Soviet environmental economists working on the project sought to establish charge schedules based on risk to human health, damage to the ecology, and lost economic productivity. Within that framework, the amount of the charge for a particular enterprise was calculated based on its actual discharges and the characteristics, such as the assimilative capacity of the ecosystem, in the area in which it was located. Table 2-6 shows the wide range of fees charged for like pollutants across regions. In the case of carbon monoxide, for example, the fees ranged from 1.2 rubles per ton in the city of Perm to 401.3 in the city of Dnepropetrovsk (Ukraine).

Initially, the overall target was to raise enough money to mitigate environmental damage in the Soviet Union. As the extent of the damage emerged, the goal became less ambitious.

As noted, the charges were to be deposited in the local environmental funds and then allocated to a variety of uses, including the purchase of pollution control equipment, ecological monitoring, and other environmentally beneficial activities. In this way regulators could kill two birds with one stone. They accomplished environmental cleanup by forcing plants to manage resources more prudently, and the system generated revenues that could be used for environmental protection. Using a charge system to control pollution made sense to many Soviet economists because of their institutional orientation toward planning and an analytic framework oriented toward centralized data collection and administration.

After several years of research, interaction with Western economists and environmental regulators, and work with environmental groups in the former Soviet Union, the economists developed and tested a series of prototype, charge-based environmental programs. Because at the time of the experiments there was no legal basis for a charge system, all the experiments were carried out under administrative fiat (decrees) or local legislation.[12]

Following is a summary of selected experiments and their outcomes.

The Kostroma *Oblast* Experiment

According to Yuri Protasov of the Kostroma Oblast Committee for Nature

---

[11] The charge program was similar to an approach developed by a group of economists at EPA in 1981-83. It involved an integrated, multi-media approach to regulating toxics that ranked environmental problems geographically based on risk. EPA staff did not carry out the analysis needed to establish fees for pollutant pathways specific to cities or regions.

[12] The earlier Law on Industrial Enterprise did allow environmental authorities to collect fees from industrial organizations, but it did not specify how to collect the money.

TABLE 2-6
CONFIRMED STANDARDS FOR FEES FOR POLLUTION OF AN AIR BASIN
(rubles/ton)

Substance	Perm	Ryazan	Ulyanovsk	Dnepropetrovsk
Sulfuric anhydride	24.00			
Carbon monoxide	1.20	56.60	44.00	401.30
Nitrogen oxide	14.10	5.10	5.00	18.20
Hydrocarbons	12.00	141.50	128.00	749.80
Sulfuric acid	12.00		20.00	23.00
Hydrogen sulfide	24.00	122.50	399.00	839.90
Ammonia	6.00	141.50		999.70
Silicic dust (>=50%)			146.00	
Metallic dust		283.00		
Nitric acid		249.00		
Solvent		9.10	21.00	

Source: Goskompriroda, "Economic Experiment on the Development of Economic Mechanisms in the Use of Natural Resources," 1991, p.22 (translated by John Metzler).

Protection, the Kostroma experiment conducted in 1989-90 validated the legal, scientific, economic, and technical nature of the fee system. Each year, 5-6 million rubles were collected and recycled into pollution control activities. While this revenue represented only a small part of the investment needed for pollution control, the funds contributed to specific cleanups, and the experiment was an important first step in the use of economic tools.

Three steps had to be taken to allow the Kostroma experiment to proceed:

(1) A data base on pollution had to be established.

(2) Each firm had to submit a so-called title page with information on its emissions, effluents, and solid waste, to serve as a model "environmental passport." The passport was a document that included an inventory of emissions, effluents, and control technologies, if any.

(3) The Kostroma government had to work out a permit with each firm stipulating that the firm would pay a charge for emissions, effluents, or disposed waste within the permissible limits and higher fees for pollution in excess of the permissible limits. That is, the charge system embodied both a user charge and a penalty system. The

permit was an official document, similar to a contract, that allowed the holder to release a certain amount of pollution.

In 1990, 1,090 enterprises and firms in the Kostroma *oblast* took part in these experiments. That level of participation was high. However, concerns about energy problems prevented Kostromotenergo and Kostromalesprom (the electricity and gas companies) from joining in the experiment. The participating firms in this experiment paid 6.1 million rubles in fees, of which 30 percent went to the local ruling councils and 70 percent to the special fund for environmental protection. Of the 4.3 million rubles dedicated to the fund, 4 million were spent on the construction of cleanup facilities, scientific research, and the acquisition of instruments and equipment.

Based on the outcome of the experiment, its managers suggested that a nature protection fund be formed and used by the committees for environmental protection projects. It was recommended that these committees coordinate their work with the regional economic committees. Clearly coordination is an important issue, and, along with it, staffing, as coordination requires a multidisciplinary, professional staff able to communicate with a variety of stakeholders. However, regional authorities did not initiate training programs for the staff.

The Nizhnii Novgorod Experiment

The Nizhnii Novgorod experiment had three goals:

(1) Improving the ecological situation in a region with serious environmental problems.

(2) Defining the methods for calculating charges and the design of the charge system.

(3) Ensuring approval and use of the economic mechanisms by the self-governing regions.

This experiment stressed water pollution and solid waste control, calculation of emission limits, and methods of calculating fines. The limits for air emissions were calculated based on three factors—risk, total emissions in the area, and the damage caused by individual contaminants. Three approaches to the charges were used:

(1) The committee determined how much was needed to achieve a certain environmental quality in an area and then collected that amount from enterprises based on a formula.

(2) Industry was required to pay for the economic damage caused by pollution, with firms required to cover their damage.

(3) With respect to water pollution, individual firms were required to pay

82   *The Environment Goes to Market*

a fee based on the volume of water necessary to dilute sewage to reach health standards.

(4) Supplementary fines were issued for air pollution and administrative violations, such as operating without an emission permit, inaccurate reporting, and late reporting. These fines were a function of emissions.

Nizhnii Novgorod's Committee on Ecology and Nature was responsible for disbursing the revenues collected. Revenues from the environmental protection fund went for technology for monitoring and control (18.4 percent) and research into the ecology (15.7 percent) and for industrial waste treatment (8.3 percent) (table 2-7). The committee also set aside reserve monies in the funds.

A Case Study in the Moscow Region

On April 14, 1990, the Moscow Executive Committee issued Decision number 840 entitled "Concerning the Experiment on the Introduction of Economic Mechanisms for Environmental Management in Moscow." The proposal was for a demonstration project involving emission charges for Moscow. The city government did not formally adopt the proposal referred to as "Concerning the Introduction of an Economic Mechanism for Environmental Management" (No. 221) until December 3, 1991. The reasons for the delay were major political and economic changes, including a reorganization of Moscow's management structure, elimination of the former district-based local governments, and establishment of new forms of local government called prefectures. Nevertheless, a demonstration project was implemented in 1990 and 1991.

The first stage of the experiment involved calculating estimates of the economic damage from environmental pollution. The value was found to be roughly 3 billion rubles. It was also estimated that enterprises in Moscow generated total profits of about 40 billion rubles. These figures were to have been the basis for calculating the pollution charges.

A number of problems required abandoning this approach. Industrial leaders rejected the concept of taxing their profits for environmental protection, and at the beginning of the experiment there was no legal means of forcing them to pay the charges. Further, the incentive mechanism embodied in the charges did not work well because enterprises had few environmental products and services available with which to lower their emissions. Finally, the drastic economic downturn of the period left many firms vulnerable to closure, which would have increased unemployment and social suffering. As a result, the authorities were reluctant to impose the charges. In short, the timing for the introduction of new pollution charges could not have been worse.

The institution primarily responsible for enforcing ecological decisions in Moscow was the Moscow Environmental Committee. At the start of the experiment, this committee adopted temporary instructions for calculating and collecting pollution charges for the Moscow Environmental Fund and for conducting environmental assessments of Moscow firms. The committee also

TABLE 2-7
TRENDS IN THE USE OF FUND REVENUES IN THE CITY OF NIZHNII NOVGOROD FOR THE PROTECTION OF NATURE

Use	Percent
Preliminary work on the development of new technologies for the protection of nature and the creation of an automatic monitoring system.	18.40
Interdisciplinary research on the protection of nature.	15.70
Formation of state enterprises researching industrial wastes on local territory.	8.30
Creation of a state fund for the protection of nature.	4.15
Creation of health protection for illnesses caused by pollution.	5.50
Awards and incentives for industrial enterprises not within the control of Goskompriroda or other social protection agencies. Industries will have direct control over the protection of nature and the rational use of natural resources.	3.70
Preliminary work on the valuation of factors influencing the environment and on the ecological aspects of economic projects, implemented on the territory of the Congress of People's Deputies.	4.50
Reserve.	16.60
Construction, military conversion, reconstruction, capital expenditures, and repairs of machinery used with the environment in the local region.	4.50
Financing of environmental measures at enterprises.	0.92
Other.	17.73
TOTAL	100.00

Source: Goskompriroda, "Economic Experiment on the Development of Economic Mechanisms in the Use of Natural Resources," 1991, pp. 47-48 (translated by John Metzler).

adopted new rules for environmental permits. Committee members visited other cities and towns that used environmental charges, including Kemerovo, St. Petersburg, Sumju, Irkutsk, and Kostroma, to learn from their experiences. The committee also launched a public information campaign, holding more than 50 lectures and seminars to introduce the experimental program, and began informing area newspapers about which firms were in gross violation of the pollution laws and how much they were supposed to pay.

At the same time, a special environmental department was established within the municipal services department. It issued guidelines for charges for sewage leaks and water purification. These guidelines had an immediate impact—they lowered total pollution from industrial enterprises by roughly 39,000 tons and sewage with heavy metal content by 166.5 tons.

Assessment and collection of the fees were and continue to be under the finance/economic inspection group of the Moscow Environmental Committee. This group collects and audits emissions figures and reviews the charges the enterprises calculate they owe. As of January 1, 1992, it had received 1,092 calculations from enterprises. Of these, it processed 961, which were put into a special index. This process is required by law, and authorities regularly carry out audits to verify that the pollution charges have officially been filed and the charges paid. Violators are prosecuted.

The Moscow government established both the pollution charge system and environmental funds to address specific environmental problems in Moscow, including the treatment of municipal waste, motor transportation, hazardous waste treatment, use of pesticides, and land management. The funds were divided into a number of smaller funds, each with a specific investment area. Some of these funds supported the Moscow water distribution and sewer system, Motorcar Inspection of Moscow, the Environmental Scientific and Business Association, and the Moscow Meteorology Committee. In 1991, the Moscow Environmental Fund received 11.2 million rubles.

Participation in the charge program grew rapidly, and there was a growing balance in the fund. The regular payments were supplemented by special charges levied against enterprises with high levels of contaminants. The courts, arbitrators and public prosecutor handled these charges. In 1991, 95 special charges were initiated.

Despite the growth in fund receipts, certain factors limited how much was collected. The Environmental Committee set the maximum level of charges for a single enterprise at 7 percent of profits. However, it exempted state-owned enterprises, along with many construction firms. Similarly, the city utility firms, including the Moscow power plants, did not have to participate in the charge system, nor did the largest employer in the region, the ZIL production association (an auto manufacturer). Had those utility firms and ZIL participated, their estimated fees of roughly 10 million rubles would have nearly doubled the receipts of the environmental fund in 1991.

Agreements were reached on the investment of 3 million rubles in environmental projects, with representatives of the Environmental Committee and Moscow government responsible for the decision. The Moscow City Council then objected to the arrangement. Legally, however, the city council had no say in the matter, because the fund, as part of the Russian Environmental Fund, was independent of the council. Currently, allocations are handled using the guidelines set forth in Moscow Executive Committee Decision 840.

In 1991, allocations from the fund went to a number of projects, including:

- Arranging an international conference on solid waste (in partnership with the MEP&NR).
- Development of city parks.
- Environmental assessment of two industrial districts in Moscow.

- Staff support for the Moscow Environmental Committee (less than 2 percent of the fund).
- Office and automation equipment for accounting, planning, and analysis of enterprise activities.

As of January 1, 1992, 3.7 million rubles had been dispersed and 7.2 million rubles remained in the fund. Because of inflation, the remaining amount has dwindled considerably in real terms. However, it is still significant. Efforts are under way to link the charge levels to inflation so as to maintain a solid balance.

It is not surprising that in its first seven quarters much of the fund was spent on assessment and institution-building. This trend could continue for many more quarters, with a gradual shift toward financing cleanup projects. Tangible cleanup efforts will not be delayed until then, however. Because the Law on Environmental Protection allows enterprises to undertake their own cleanup in lieu of paying into the fund, many firms opt to keep decision-making and finances in-house and carry out their own cleanups.

Because of legal changes in 1992, continued evasion by enterprises that had avoided paying the charges became more difficult than in the past. By the time the Russian economy starts growing, there will be even fewer ways to avoid payment. The Moscow fund should swell dramatically. Moreover, the cooperation between the Moscow Environmental Fund and other governing groups such as the local environmental protection agency is a positive sign, because the cost of many ecological projects of great importance (such as advanced municipal waste treatment facilities) requires a consortium of public entities.

The credible threat of prosecution is crucial to the Moscow environmental regulatory program. Investigation and prosecution of violators during the experiment demonstrated at least a minimum capacity for enforcement. As noted, the changes in environmental laws should make the threat of prosecution even more credible.

Initiating a National Program

In January 1991, the Russian government issued a special regulation, Resolution 13, that introduced the pollution charge program throughout the Russian Federation. In developing the resolution, the government drew on the experience of the various regional experiments. The resolution, which covered emissions from both stationary and mobile sources, defined a system of required payments that would go to special funds. The fees varied according to the type of pollution and the characteristics of the locality, in contrast with uniform fees. Resolution 13 and supplementary regulations were needed to strengthen the legal basis and enforceability of the charge system, given the many shortcomings in the existing set of laws.

In December 1991, Russia became a sovereign state, effectively breaking up the USSR. Already in the fall of that year, Prime Minister Yegor Gaidar had formed his cabinet of "economic reformers" and strongly supported the notion of a market economy. Although the cabinet rescinded or revised most previous economic decisions, it extended the 1991 pollution charge program into 1992. On December 19, 1991, the Russian parliament adopted a new Russian Federation Law on Environmental Protection (EPL), which President Boris Yeltsin signed on February 19, 1992. Finally there was a legal basis for the pollution charges and non-budgetary environmental funds. Since the beginning of 1993, the government has made minor changes to the program, but its main features remain in place.[13] The EPL was part of a larger effort to create a framework of environmental legislation and regulation in Russia. For example, subsequent legislation established penalties for the misuse of water, land, and timber. The EPL should be seen in this larger context.

OVERVIEW OF THE POLLUTION CHARGE SYSTEM

As context for the overview of the pollution charge system, this section first provides some background information on Russia's environmental and economic situation, the basic principles of Russian environmental policy, and the organizational structure within which the system operates. It then reviews attitudes toward the pollution charge system, the main features of the charge system, the method of calculating the charges, the collection of charges, and the allocation of revenues.

Contextual Background for the Pollution Charge System

The Environmental Effects of Industrial Growth

To understand Russia's current environmental legislation and regulatory programs, seven important characteristics of its environmental and economic situation must be understood:

(1) *Russia's economic crisis.* The federal budget is running a substantial deficit, and the majority of what revenues there are go to state-owned enterprises, whose production levels are off drastically.

(2) *The monopolistic nature of Russia's economy.* There is still only one major supplier of most industrial goods, a condition that inhibits regulators from imposing heavy sanctions on enterprises whose production is essential to other industries.

(3) *The severe human health problems in many regions.* Russia ranks 51st in the world for average life expectancy. Environmental factors (mainly air pollution) are estimated to account for 20-30 percent of the causes

---

[13] Of the Eastern European countries, Poland and Czechoslovakia also initiated pollution charge programs before their transition to a market economy, and they, too, have continued them, with modifications. Based on the experience of the Eastern European countries, pollution fees appear to be an adequate instrument of environmental policy for economies in transition.

of death. Environmental conditions in urbanized areas and industrial regions are appalling, in some cases exceeding the health standards by factors ranging from 10 to many 100s. Last year, the environmental standards for health were exceeded by a factor of 10 in 84 cities. Nearly half the pipelines for drinking water are health hazards because of the poor condition of municipal and industrial water treatment systems. Recent studies showed high levels of dioxin, benzopyrene, and polychlorinated biphenyls, the most dangerous pollutants. The concentration of polychlorinated biphenyls in selected tests of breast milk in Moscow, Baikalsk, Postov and some other cities was 5-12 times higher than the level permitted in the United States.

(4) *Unsatisfactory pollution control devices.* More than half of all industrial enterprises have unsatisfactory pollution controls. Historically, industry did not regulate its emissions of pollutants, effluents, or hazardous waste. Many did not even implement cost-efficiency measures, such as energy-saving devices, that coincidentally would have reduced emissions. Most of those firms that have end-of-pipe controls do not operate or maintain the systems in accordance with design standards.

(5) *The small size of the Russian environmental goods and services sector.* Russia has had an environmental goods and services sector for 20 years now. However, it is small and quiescent, absent a domestic market, and has not been producing essential items such as pollution control and metering devices. At the same time, products imported from Western Europe, North America, and Asia are beyond the financial reach of virtually all Russian enterprises. Although a domestic environmental industry is now beginning to grow, for financial, technical, and regulatory reasons, it is difficult for the typical factory to obtain the technologies.

(6) *Inadequate government monitoring of pollution.* The previous two points also have regulatory implications. The quality and frequency of measurement and sampling of industrial emissions, effluents, and solid wastes have been low. Government sampling occurs not more than once a year and as infrequently as once every five to eight years. There is a lack not only of high-quality testing equipment but also of measurement and sampling standards. Unlike the industrialized nations (exclusive of Eastern Europe), which have established measurement protocols that allow environmental industries to develop internationally marketable measurement products, Russia has lagged behind in setting such standards. Therefore, monitoring data are flawed, and cost comparison is impossible given the lack of agreements on measurement techniques and sampling standards.

(7) *The limited scope of Russia's environmental legislation and regulation.* Not surprisingly, Russia's environmental legislation and regulation are more limited than those of other industrial nations. The situation

puts Russia at a disadvantage in developing a regulatory and private sector, compliance-oriented infrastructure.

The economic crisis has slowed progress on all fronts of environmental protection. Supporters of the green movement are less active, having to weigh economic losses against ecological gains. Governments are enforcing the environmental laws even less rigorously. Government resources are being stretched across a number of severe domestic and international problems. The weak economy, combined with the lack of measurement and sampling standards and a domestic industry that manufactures environmental protection equipment, limits the introduction of environmental technologies. This problem is especially true for wastewater treatment.

That absence of standards combines with the monopolistic character of the economy to limit the options of environmental regulators. Without acceptable standards, they have a harder time proving abuses. Further, they cannot consider drastic solutions such as closing a dangerous enterprise.

In the case of the pollution charge program, the lack of pollution control equipment and insufficient environmental legislation and regulation limit the potential benefits of recent experiments. It is possible, now that the program is officially mandated for the entire nation, that the legal uncertainties may diminish as case law and administrative precedents grow. It is more likely, however, that the tough economic times will inhibit any tightening of the legal and regulatory uncertainties and therefore leave industry even more flexibility.

The Basic Principles of Environmental Policy in Russia

The EPL established the basic framework for a new system of environmental management in Russia. The main principles underlying this law, partially detailed in government regulations in 1992, are as follows. *First,* government environmental policy is based on the ambient standards and corresponding limits for emissions, effluents, and solid waste. Existing laws and regulations establish penalties, financial liabilities, and, in some cases, criminal liabilities for violating the limits.

Two basic types of ambient standards and discharge limits have been adopted. The first type is the maximum permitted concentrations (MPCs)—the ambient standards—of pollutants in environmental media (air, water, soil, and foodstuffs), and the corresponding maximum permitted discharges (MPDs)—the source limits—for each specific pollutant (or source of pollution) for a given period. The standards and limits are based on a goal of zero damage to human health from pollution; on rare occasions, impacts on ecosystems are also taken into account.

The second type is the temporary permitted concentrations (TPCs) and corresponding temporary permitted discharges (TPDs). They are supposed to be intermediate steps toward attaining the MPCs and MPDs and are based on current technological and economic resources for environmental control.

These standards and limits are the foundation for all environmental regulation. They cover official requirements and standards for production, construction, and transportation technologies, and different kinds of commercial and non-commercial activities. All sources of pollution, stationary and mobile, entering any environmental media are affected. No industry is exempt, including military enterprises.

The limits set by the MPDs, essential for current environmental policy, are still underdeveloped. In many cases, the TPD values, which are based on vague concepts of technological feasibility, are being used instead, and even the TPD limits, are currently available for only a limited number of pollution sources.

Second, development and implementation of national environmental policy are the prerogative of the MEP&NR. The ministry and its regional offices are responsible for establishing the ambient standards, discharge limits, and other regulations, as well as for setting the base rates for the pollution charges. As noted, however, local governments can waive the charges for any enterprise.

Third, environmental pollution and natural resource use are subject to taxation. Revenues from the pollution charges are to be used to improve environmental quality. Although the EPL allocates budgetary resources for expenditures on environmental protection, the non-budgetary environmental funds, all of whose revenue comes from pollution charges, are considered the principal source of monies for activities to promote environmental quality. This is a unique and powerful aspect of the discharge fee system.

The introduction of pollution charges and non-budgetary environmental funds is undoubtedly the most important innovation of the EPL. These charges, together with value-added taxes and taxes on the use of natural resources, were practically unknown in the USSR. Now they are the main features of the new tax system adopted by the Russian Federation at the beginning of 1992.

The Organizational Structure

The pollution charge system functions within a complex organizational structure. At the top are the two chambers of the Russian parliament. The parliament establishes the legal basis for environmental policy and adopts national programs, within the constraints imposed by international conventions and other agreements on pollution levels to which Russia is a signatory. As an elected body, the parliament is influenced not just by government experts but also by public organizations such as Ecology and Peace and the Socio-ecological Union.

The executive branch of the federal government—the Cabinet of ministers—is responsible for implementing the programs established by the legislature and for seeing they are carried out according to the law. At the head of the executive branch is the president of Russia, who has a presidential advisor on environment and public health. The advisor's staff prepares draft texts of individual presidential decrees on environmental issues.

Because the federal government is still deeply involved in industrial production and decision-making, there are many industry-related ministries that coordinate many industrial activities. The federal government also decides the order in which environmental standards, including levels of emissions, effluents and hazardous waste generation and disposal, are developed and adopted. At the same time that the government is constrained by international conventions and agreements, it also benefits from international experience and cooperation.

Of the ministries, the most important, as noted, is the MEP&NR. It is responsible for environmental policy-making, and it promulgates the standards and both rules and regulations for environmental protection. The MEP&NR develops and implements the pollution charge program with the participation of the Cabinet of Ministers, Ministry of Economy, Ministry of Finance, State Tax Service, State Bank, and Committee on Hydrometeorology and Environmental Monitoring. MEP&NR's responsibilities include development of the regulations for collecting revenues and for expenditures, inspections to determine whether the charges established are correct, and enforcement.

Management of the pollution charge program at the federal level is as follows. The Cabinet of Ministers of the Russian Federation establishes a method for determining the pollution charges and their maximum amounts. The MEP&NR, the Ministry of Economy, and the Ministry of Finance then set the base charge rates and correction coefficients. The Cabinet of Ministers, in turn approves the maximum allowable charges for different industries. The MEP&NR, Ministry of Finance, State Tax Service, and State Bank together propose, formulate, and approve the regulations and instructions on how to collect pollution charges.

The ministry has a total staff of about 22,500. About 650 are at its headquarters and about 15,000 at its territorial offices. The ministry has four major departments, which specialize in different types of economic instruments for environmental protection: pollution charges; natural resource charges; ecobusiness and market development of environmental services; and finance and administration of natural resources and environmental protection. The ministry's territorial offices consist of committees and inspectorates.

The next level of the organizational structure is the administrative units that constitute the Federation: the 22 republics, 11 *okrugs*, 6 *krais*, and 49 *oblasts*. These entities are responsible for incorporating local environmental, cultural, and administrative factors into their environmental regulations. In addition, two metropolitan cities—Moscow and St. Petersburg—have the same administrative status as other federation units. Together, the units make up the regional level of management. (See figure 2-3.)

The next level down in the organizational structure is the local governments of cities, towns, and districts. Under the MEP&NR are some 200 locally based offices including 90 regional offices. The rest consist of local environmental committees in regional capitals, other large cities, and some districts. Regional offices (called ministries in the republics and committees in the other entities) are the heart of the Russian environmental management structure.

The central ministry appoints their chairmen, and the regional governments approve them. Staff levels usually vary from 200 to 400, with a significant portion working with the local environmental committees of the cities and rural districts. As a rule, these offices include the following departments: economic instruments for environmental and resource management; environmental reviews of economic projects; environmental inspection; and natural resource public records. Some regional offices have taken steps to cooperate and have formed inter-regional inspection teams. The regional offices report to the central ministry offices in Moscow, primarily through its Department of Coordination of Regional Bodies Activities.

The central ministry and other outside institutes provide the committees with guidance on policy, rule-making, and resource allocation. The committees are also influenced by local environmental groups, especially in the case of decisions on where to locate industry. The regional and local governments, with the participation of and in cooperation with the regional and local environmental committees, play a key role in the charge system:

FIGURE 2-3
ORGANIZATIONAL STRUCTURE OF ENVIRONMENTAL MANAGEMENT IN THE TERRITORIES OF THE RUSSIAN FEDERATION

Federal Level

Russian Federation
Central Office of MEP&NR

Regional Level

Republic (22)	Okrug (11)	Krai (6)	Oblast (49)	The cities of Moscow and St. Petersburg
Republican Offices of MEP&NR (Ministries)	Regional Offices of MEP&NR (Environmental Committees)			

Local Levels

Cities, Towns, Districts
Local Environmental Committees

Source: National Academy of Public Administration.

- They issue the permits that allow local enterprises to use natural resources.

- They make the final decisions on both the rates that enterprises have to pay for natural resources and the levels of the pollution charges to enterprises, including exemptions of individual pollutants from charges. That is, the regional and local authorities have wide discretion to redistribute the burden of the charges among polluters, based on each one's circumstances.

- They administer the regulations on pollution charges, at the regional and local levels.

- They collect and manage the regional environmental funds. The funds' revenues come mainly from the pollution charges enterprises pay.

- They are responsible for the compliance of all local enterprises with the regulations. They work with local enterprises to determine the levels of the pollution charges and calculate or measure the actual amount of pollution. The law empowers the regional committees to obtain court orders to enforce environmental rules and collect charges.

- They provide feedback to the ministry on the results of regulatory experiments.

The high degree of autonomy the regional and local governments have to set the charge rates, collect the charges, and spend the revenues is a striking feature of the Russian system and one that sets it apart from systems elsewhere. Essentially the committees are not subject to any central oversight on what they do with the revenues.

The hierarchical nature of Russia's industrial structure means that Russian ministries own and operate many enterprises. Thus, for a significant share of Russian industry, negotiations over the level of pollution charges and other matters take place at the ministerial level. The result is negotiated between the regional and local committees, which have substantial legal power, and the ministries, which have substantial economic power. The result of the negotiation is not always predictable.

The final set of organizations involved in environmental legislation and regulation in Russia are the enterprises themselves. They are responsible for paying the pollution fees and charges for the use of natural resources, as specified in their permits and by the ministry and local authorities. They are also responsible for changing their production processes to lower their emissions of pollutants and, in turn, their payments. The law allows them to spend internal funds on environmental protection measures; with local approval they can deduct those costs from the required pollution charges. Finally, the enterprises try in various ways to influence the legislative and regulatory processes, including through sympathetic political parties.

Enterprises are required to report their own pollution levels. The local agencies review the data submitted and determine whether inspections are needed. They may inspect an enterprise regularly or rarely. As with all self-reporting systems, it is possible some enterprises underreport their pollution levels. To date, the authorities have uncovered and pursued several hundred violations, many of which have resulted in punishments. However, in general, governments at all levels do very little monitoring of pollution levels, and enforcement is therefore lax.

In other countries there is a further level of participants—organized environmental pressure groups. The absence of such groups at all levels of decision-making and action in Russia is striking. Currently, organized input from environmental interests occurs only at a very high level (the parliament), with sporadic input on single issues at the local level, usually related to the siting of facilities. For now, the pro-environmental stance of almost all local governments and committees in Russia compensates for the lack of formalized environmental input. As the system evolves, however, it is conceivable some local governments will drop that pro-environmental stance, and environmental interest groups will become necessary.

ATTITUDES TOWARD THE POLLUTION CHARGE SYSTEM

After years of ignorance and refusal to acknowledge the problem, officials now generally support pollution charges as well as other economic instruments for pollution control. The MEP&NR staff, particularly those in the economic instruments departments at the central and territorial offices, view the charge program as a positive innovation; some even consider it the most efficient instrument the ministry has. They have become an influential group that actively supports further development of economic incentives for environmental management.

If officials at the central ministry stand for centralized decision-making in applying economic instruments, regional and local environmental administrators are interested in more decentralization. Some of the local environmental administrators—who are younger, more active, better educated, and not entrenched in old habits of management—are looking for new economic incentives for environmental protection. For example, they are experimenting with marketable permits. Their ideas are meeting with increasing, albeit anxious, interest among progressive, market-oriented industrial leaders. Clearly, however, there is tension between the centralized ministry staff and the regional desires for autonomy.

The position of the industrial circles (consisting mainly of the directors of large plants, who in the old days were often appointed by the Soviet government) and of the green movement is hard to pinpoint. Within both the management elite and among the "greens" are many people who still are nostalgic for Soviet ideology and the past and are highly skeptical of a market economy in general and economic incentives for environmental protection in particular. The rest, who are progressive, are better disposed toward pollution charges and market-oriented solutions in general, although even they are prone to pointing out the failures and not the successes.

As to local enterprises, historically their decision-making has been vertically oriented, given that powerful ministries in Moscow have been managing them. In the past, these enterprises counted on their central ministries to clear the way for their industrial production by dealing with environmental and other issues. Today, the once-mighty central ministries cannot provide the same level of political or resource support. As a result, more than ever enterprises are having to deal with local authorities. Consequently, many are sponsoring or supporting sympathetic candidates for public office at the local and national levels whose positions may be in conflict with stressing environmental cleanup.

In the meantime, the MEP&NR is sure the new EPL is influencing the behavior of enterprises. Its staff report changes in industrial strategy to reduce the charge levels in areas where the program has been in effect for 12-18 months. Enterprises are developing their own, self-funded environmental protection projects. Historically, the entire environmental budget came from the central authority—about 2 billion rubles a year. Now enterprises are expected to spend roughly the same amount on their own projects (with inflationary corrections).

The green movement in Russia does not participate in policy-making as much or in the same manner as it does in the United States. There are green members of parliament, and the public writes letters supporting environmental causes to members of the government. Two major environmental newspapers also address these issues. Currently, however, the economy has left the green movement weak, and as an organized movement, it is not very large. In regions where recognized problems exist, such as Chernobyl and Baikal, the greens have organized, and local groups have also come together to oppose the construction of individual plants.

The green groups do not have sophisticated mechanisms for generating financial support and generally do not have the expertise to follow environmental regulation through the system in detail. Environmentalists have no designated seat at the table when it comes to negotiating the levels of charges and actual payments.

So far, the pollution charge program has not been weakened by this situation, because at the local level most elected governments and environmental committees are strongly pro-environment and effectively represent the green viewpoint in public matters. Efforts are under way to improve the organization and effectiveness of environmental groups throughout Russia. As of now, however, there is nowhere near the level of institutionalization of environmental perspectives found in the United States.

To the extent the greens are mobilized, their principal intention is to sue environmental violators in an effort to penalize them through the courts or arbitration. Although the EPL permits compensatory damage through environmental lawsuits, the government has not yet approved regulations and instructions in the area. This is one reason it is relatively rare for environmental cases to go to court. Two other reasons are the absence of a tradition of using the courts for non-criminal matters and the lack of qualified attorneys. In addition, the courts are clogged with civil cases. The consensus of all stakeholders is that environmental regulations and enforcement actions need to become more efficient in use and somewhat predictable in outcome.

The media do not play a significant role in environmental issues in Russia. They have shown little interest in the issue of pollution charges and have paid it little attention. The exception is the environmental publications, especially the two main green newspapers, *Zelyonyi Mir (Green World)* and *Spasenie (Salvation)*. These publications routinely print articles explaining the advantages of economic instruments. On the other hand, a reader is not likely to find articles analyzing the practical problems of the charge program.

As for the public at large, pollution charges have not generated much interest. Generally, surveys of public opinion reveal that environmental problems are less an issue now than they were at the beginning of *perestroika*, when the green movement reached its zenith. Social instability, inflation, and the general fall in economic welfare account for this loss in interest. Nevertheless, the public still rates the environment as a high priority, particularly in regions where there have been environmental disasters.

Main Features of the Pollution Charge Program

In 1991-92, Russia adopted pollution charges for air emissions, water effluents, and waste disposal. In 1991, MPC standards applied to 479 air pollutants, 2,675 water pollutants (including 1,050 rates for fish-breeding water basins), and 109 soil-polluting substances. TPC standards applied to 1,138 air pollutants and 69 soil pollutants. The plan is to introduce charges for noise, electromagnetic, and some other kinds of environmental pollution in 1993. Following are the main features of the current system as it stood in 1993, after some minor changes.

- Three categories of pollution charges apply to discharges, based on whether they are (1) *within* the MPD; (2) *in excess of* the MPD but *within* the TPD; or (3) *in excess of* the TPD. The base penalty rates for a pollutant in the second group of discharges are up to 25 times higher than the group 1 rate for the same pollutant, and the group 3 rates are five times as high as the group 2 rates. The latter rate is imposed on overall discharges if a polluter does not have either a TPD or MPD discharge limit.

- Payments for group 1 charges are treated as a cost of production, whereas the group 2 and 3 charges must be paid out of net profits. However, the government has established maximum levels for group 2 and 3 charges paid by different industries. If a polluter's profit after taxation is equal to or less than the total group 2 and 3 charges, the government may halt the firm's operations temporarily or permanently.

- The total amount of the pollution charge each polluter pays depends on five factors:
  — The types of pollutants discharged.
  — The volume of the pollution discharges within and above thepermitted levels.
  — The base charge rate in rubles that the MEP&NR has established per ton (cubic meter) of polluting substances and solid waste. As of 1993, the charges were set jointly by the MEP&NR, the Ministry of Economy, and the Ministry of Finance.
  — Correction coefficients applied to take into account the particular environmental and socioeconomic conditions in each region and the corresponding potential damage caused by the pollution. These coefficients are calculated centrally (usually in terms of maximum/minimum figures). However, the regional authorities have the right to decide whether to apply them or how to modify them.
  — Privileges and exemptions given to a specific polluter by the regional or local authorities related to the polluter's expenditures for environmental protection.

- The pollution charges are distributed to the federal budget (10 percent) and the network of federal non-budgetary environmental funds (90 percent). The latter amount is distributed across the federal level (10 percent), regional level (30 percent), and local level (60 percent).

- The pollution charges do not exempt polluters from their obligation to carry out environmental protection measures and to pay compensation for environmental and health damage resulting from environmental violations.

The 1993 pollution charge system differs from the system in effect in 1991-92 in three ways. First, the former system used two instead of three types of pollution charges: (1) it taxed permitted amounts of discharges, generally based on the TPD and rarely on the MPD quotas; and (2) it levied penalties for emissions above the permitted quotas. Second, under the former system, polluters paid both charges out of profits after taxation. Third, previous regulations gave regional and local authorities the right to reduce or waive an enterprise's entire charge in the event of difficult economic circumstances. The new regulations permit waivers only to facilitate an enterprise's investment in the control or treatment of pollution.

The rest of this analysis focuses on the system in place before 1993 because comprehensive information on the revised system is not available. However, the principles and problems of both systems are similar.

Calculation of the Charge Rates

The charge program specifies nearly 300 base rates for air pollutants and nearly 150 for water pollutants. The method used to calculate a base rate charge is as follows. The amount of every pollutant is measured in equivalent pollution tons (EPTs), which are equal to the amount of the pollutant in metric tons divided by the MPC values. The MPC values are assumed to reflect the relative health risk to humans caused by the corresponding pollutants. As a result, the MPC rates for different substances vary widely. This diversity is automatically reflected in the EPT and therefore in the actual pollution charge rates. For example, the air pollution charge is 0.7 rubles per ton for hexane and 6.6 million rubles per ton for tellurium dioxin (for other examples see table 2-8).

Charges also are calculated based on the prior estimated amount of charge receipts. The central government establishes this amount at a level that is considered "acceptable" for polluters but that is also sufficient to build up the environmental funds. All corresponding assessments of charges are, however, necessarily vague.

TABLE 2-8
CHARGE RATES FOR EMISSIONS, 1991
(rubles/ton)

Pollutants	For emissions within the limit	For emissions over the limit
Nitrogen oxide	90.75	434.50
Sulfur oxide	66.00	316.00
Carbon oxide	1.09	5.21
Dust (with siliceous component > 20%)	33.00	158.00
Methane	44.00	235.00
Benzopyrene	1,099,998.00	5,266,661.40

Source: Supplement to the Council of Ministers of Russia Federation Regulation Number 13, January 1991.

The government has not developed strict rules on how to spread the financial burden for environmental protection across the federal, regional, and local budgets; non-budgetary environmental funds (national, regional, and municipal); and polluters and consumers. The central authorities responsible for setting the pollution charges have none of the data, such as the marginal costs of pollution abatement and the economic cost of damage resulting from pollution, required to make even a rough estimate of the level of charges that would be a sufficient incentive to curb pollution.

This system of calculating charges is excessively complex, precise, and centralized. These characteristics are only partially neutralized by the delegation of broad discretion to local authorities to correct the recommended rates and rules for implementation. Nevertheless, despite the uncertainty embodied in this approach, the preliminary results associated with implementation of the charge system are perceived to be largely positive.

Among the immediate problems is that of inflation. When the rate of inflation is high, it depreciates the receipts from pollution charges based on units of pollution. This is the situation in Russia. To offset the high rate of inflation, during the second part of 1992 the government increased the initial pollution charge rates 500 percent. However, in the face of an inflation rate estimated in 1992 to be nearly 2,000 percent, that increase was insufficient.

Collecting the Charges

Enterprises are invoiced quarterly, no later than the 23rd of the month. If the deadline for payment passes, money is transferred automatically from a polluter's account through a "payment demand" form presented directly to the bank where the defaulting polluter has an account. Polluting enterprises have to honor the bank's "payment order" regardless of the amount of money in their accounts. If a polluter is temporarily insolvent, the bank transfers the payments into the environmental fund as soon as money appears in the polluter's account.

As noted, when the charges for pollution are equal to or greater than the profit left after the enterprise is taxed, the territorial offices of the MEP&NR, the Committee on Hydrometeorology, and the Sanitary-Epidemiological Inspection Unit must notify a regional or local government to request a stop to the enterprise's operations. The government is not, however, obligated to grant this request.

Allocating the Revenue from the Charges

Initially, all the receipts from the charges go to the regional environmental funds at the republic, *krai*, *oblast*, and *okrug* levels and to Moscow and St. Petersburg. From there, 10 percent is distributed to the federal budget; of the balance, 10 percent goes to the federal environmental fund, 30 percent is retained at the regional level, and 60 percent goes to local funds.

As noted, the EPL and related regulations give the local funds wide discretion in allocating revenues. According to the *National Report of Russia on the Environmental Situation* in the Russian Federation in 1991, the revenues from the environmental funds were distributed as follows: 86.3 percent for research on and design of low-polluting and resource-saving technologies and equipment; 3.6 percent to build and maintain reserves, preserves, and natural parks; 3.6 percent to environmental impact statements and environmental reviews; 2.7 percent to education and information; 1.3 percent to develop environmental standards; 0.8 percent to an environmental information system; 0.8 percent to restore, reproduce, and protect natural resources damaged by economic activities; 0.8 percent to international cooperation; and 0.1 percent to help in the design of republic-level programs for environmental protection.

Final Observations

In Russia, the procedure for collecting charges differs greatly from the official one. One reason is that over the last few years most enterprises became chronic defaulters on their obligations. A great many firms were not paying the taxes on their profits and property or the pollution fees. There have, however, been few cases of bankruptcy, despite presidential and governmental decrees establishing procedures for it. Bankruptcy, although common in a market economy, has not existed in Russia. As such, non-payment of pollution charges does not carry the threat it does in the West.

The government's inability to collect the pollution charges is the main reason actual revenues in the environmental funds have fallen short of estimates. The revenues to be derived from the charge program in 1991 were targeted at 4.6 billion rubles. Only 2.5 billion rubles were collected (this amount includes 1.0 billion spent by the enterprises for environmental protection). At the mid-point of 1992, the targeted amount for total revenues in the funds was 6-7 billion rubles. According to official government estimates, only about 2.5-3.0 billion rubles had been collected. For 1992 as a whole, the amount collected was less than half the estimated amount, despite the 500 percent increase in the charge rates between September and December. The revenues obtained by the funds in that year were less than what the Soviet Union had spent on environmental protection and centralized investments five years earlier in 1988—when total expenditures were about 7.0 billion rubles, or about 1.3 percent of gross national product (GNP). (The comparable figure for the United States was 2 percent.) Of all the taxes collected from enterprises and organizations in Russia, the pollution fees amounted to less than 0.2 percent. Even when monies for environmental protection from other sources are added to those in the environmental funds, a severe shortfall in financial resources is obvious.

According to specialists, two factors (in addition to a delay in passing the bankruptcy law) accounted for the failure to collect what had been targeted. First, using the rights the federal law delegated them, regional and local governments often lowered the charges to polluters. It appears the polluters and regional authorities had an agreement whereby the latter would not exacerbate the difficult financial status of an enterprise by collecting the pollution charge in full. Officials, when asked, did not identify the criteria they applied in modifying the charges. Factors such as economic and risk-related assessments do not appear to have been taken into account in any systematic way.

Second, the regional authorities have other rules that offer additional ways to reduce the charges, such as deducting an enterprise's investments in pollution control or other costs from the charges.

THE ENVIRONMENTAL FUNDS

Within the context of the overall charge program, this section looks at the environmental funds in greater detail, in particular at the regulations governing the spending of fund revenues and management of the funds.

Regulations Governing the Spending of Revenues

According to a regulation issued by MEP&NR minister Victor Danilov-Daniliyan in June 1992, the environmental funds are to spend their revenues to:

- Implement regional and inter-regional projects aimed at improving the environment and human health.

- Develop and carry out regional (republic, *okrug*, *krai*, and *oblast* and local (city and district) programs for environmental protection and restoration of natural resources.

- Conduct research and design projects in the areas of environmental control, treatment and cleanup, valuation of natural resources, economic instruments for environmental and resource management, low-polluting technologies and environment-friendly equipment.

- Support enterprises, research and development organizations, and individuals that introduce environmentally sound technologies.

- Give loans and subsidies to enterprises for the construction, reconstruction, modernization, and capital repair of control facilities.

- Support the design of computer systems for environmental monitoring.

- Construct or share in the construction of treatment and other protective facilities.

- Maintain *zapovedniki* (wild nature reserves), national parks, and other areas with special protected status.

- Provide economic incentives for the development of an environmental services market and environmentally sound economic management.

- Share in the costs of maintaining natural resources.

- Develop environmental and natural resource regulations.

- Raise additional money for the funds through various activities.

Marketable Permits and Pollution Charges

- Develop international cooperation, including participation by foreign experts and organizations in consulting services, environmental reviews, and other work.
- Compensate people for the damage to their health caused by pollution.
- Promote environmental education and enlightenment.
- Pay bonuses to the staff of individual enterprises and MEP&NR if they are successful in environmental protection.
- Provide benefits for the staffs of the environmental committees.
- Pay the salaries of the administrative staff of the environmental funds.
- Form the Federal Environmental Fund.
- Share in providing the regional and local environmental committees with facilities, equipment, materials, and other items.
- Carry out environmentally related projects if they are in compliance with the legislation of the Russian Federation.

In short, using the environmental funds for anything other than environmental protection or resource conservation is prohibited. The territorial offices of the MEP&NR and administrators of the corresponding funds are responsible for ensuring that earmarked expenditures are appropriately allocated.

Money from the environmental funds is spent in line with approved expenditure targets. The funds go for subsidies for enterprises and organizations so that they can complete environmentally beneficial projects on the basis of agreements between the enterprises and the regional environmental committees or fund directors. Money left in the funds at the end of the financial year can be carried over to the next year.

Fund Management

As specified in the EPL, the cabinet determines how to establish and allocate the Federal Environmental Fund, which was set up in June 1992. Decisions within the framework of government-approved expenditure targets are made by the administration of the fund. The administrative staff of the Federal Environmental Fund works closely the MEP&NR. At the same time, the fund is somewhat independent with respect to decision-making and the allocation of revenues.

Similar administration of the environmental funds exists at the regional and local levels. The regional governments supervise their corresponding funds through environmental committees or the regional fund administration, if the latter is independent from the committees.

ADMINISTRATIVE ASPECTS OF THE PROGRAM

This section addresses the issues of monitoring, enforcement, personnel training, oversight of the implementation of pollution charge systems, and program financing.

Monitoring

A key factor experts commonly consider in deciding whether to apply pollution charges is the feasibility of monitoring. A distinguishing feature of the Russian charge program is the absence of an appropriate monitoring system. A major reason is the lack of monitoring equipment and personnel. The enterprises, which are, as noted, responsible for monitoring their own discharges and types of pollutants, have more than the regional and local environmental monitoring agencies have, but what they have is still insufficient for the task. Complicating the situation is the number of pollutants subject to charges. The government is able to nationally monitor only a few of the common and voluminous ones.

In most cases, the amounts of emissions subject to charges are determined not on the basis of measurements but rather with the help of so-called material balances of enterprises. This method takes into account the characteristics of the fuel and raw materials as well as technological performance of production and pollution control equipment. The problem is that the people who choose the technologies for both production and pollution control are also the ones who calculate the emissions. Because of their reliance on design parameters instead of actual conditions, their calculations practically never match actual conditions. While their environmental protection service staff calculate the emissions on special forms, the lack of enforcement staff in local regulatory agencies means that inspections are rare.

In 1991-92, the MEP&NR territorial offices were responsible for inspections. At the end of 1992, the Committee on Hydrometeorology and Environmental Monitoring was made independent of the MEP&NR. This separation of duties should produce better results. In general, the current monitoring system can be described as a mix of self-reporting and periodic inspections.

Recently the lack of financial resources available to enterprises for environmental protection has become a critical issue. The enterprises need both modern equipment and well-trained personnel. Clearly, the scale of investment and deployment in both areas is insufficient to achieve good environmental outcomes.

Enforcement

As noted, there is a general lack of an enforcement capability with respect to pollution charges. Specific problems are:

- Uncertainty and contradictions in the law and regulations on the procedures for collecting charges.

- The rudimentary nature of the charge inspection system.

- The absence of appropriate monitoring and data base systems. The lack of monitoring equipment and tools and well-organized data bases is one reason that, even where violations of the environmental law are clear, the obstacles to identifying and punishing violators are great.

The regional environmental committees are in charge of auditing how the environmental funds spend their money. The committees keep a list of those enterprises paying charges and the amount collected, receive information and calculations of charges from those paying them, and check whether the information received is correct and the full amount has been paid. Every six months the managers of the environmental funds report to the MEP&NR on their allocations. The regional and local environmental committees perform audits to ensure the environmental funds are being used as intended.

Institutional structures for auditing the collection and expenditure of revenues are now being established. In the MEP&NR territorial offices, the auditing staff usually ranges from 2-3 people to as many as 10-12 in the major industrial areas. There are not, however, enough qualified personnel. In an effort to alleviate this problem, the tax inspection arm of the Ministry of Finance has taken steps to integrate the field activities of the environmental committees.

Enforcement of legal obligations has historically been far less developed in Russia than in the United States. The world of lawyers and lawsuits, which affords both a permanent threat and source of support for American business people, plays a modest role in the Russian business sector. Few Russian lawyers specialize in liability issues, and many Russians still associate enforcement with criminal law. Within the courts, property rights and liability issues occur mainly in connection with divorce cases. The courts are not ready to face the stream of "corporate" cases common in a market economy. In particular, the law in Russia associated with environmental damage is embryonic. Absent the threat of litigation and damages, some businesspeople will always treat the environment as a free good, and both command-and-control and market incentive systems will fail.

Personnel Training

The staff of the central office of the MEP&NR is generally well educated—more than 95 percent hold university degrees. However, the staff's strength is in scientific fields, with a shortage of expertise in the legal, economic, and policy analysis areas. At the territorial level, the average educational levels and qualifications of staff are far less impressive. The same situation is found within enterprises themselves, which at present have the greatest responsibility for environmental compliance.

With respect to staff training, the ministry does not have a formal program of its own. Instead, in the 1991-92 academic year, 590 MEP&NR employees country-wide took courses at universities to improve their qualifications and skills, and 20 in Moscow received computer training. A fairly large number of the inspectors in the MEP&NR and the Ministry of Finance also participate in training activities. On the other hand, training is generally available to only a small portion of the environmental protection and financial service staff of enterprises.

The environmental committees and enterprises similarly lack well-trained staff, a serious weakness of the charge program. To improve the situation, the pollution charge department of the MEP&NR and some regional offices (the larger ones with relatively higher levels of skills) have been preparing and issuing instructions for charge program users. However, the instructions neither explain the idea of the charge program, nor analyze the practical problems associated with its implementation. Moreover, they are rarely available to users at polluting enterprises, remaining largely in the hands of the inspection services.

Personnel who work on the charge program receive some training at seminars and meetings held on Economic Instruments for Environmental Management in Moscow or in the regions. Several short-term college and institute courses offer modest training associated with pollution charges. However, the training does not focus systematically on the practical issues staff will confront in the field.

Oversight System and Evaluation Criteria

The Pollution Charge Department of the MEP&NR is formally charged with collecting and analyzing information on program implementation and making recommendations. However, the oversight system for the pollution charge program is rudimentary. There are no specific procedures and criteria for evaluating the program. The sociopsychological and institutional aspects of the program's start-up and implementation, such as reaction by industry, the green movement, local authorities and other interest groups, consensus-building among these groups, and problems with staff training, have not been studied, nor have methods and approaches for conducting the studies been elaborated.

As for evaluation, the criteria for official evaluation are said to be based on the two main purposes of the pollution program:

- To provide economic incentives for polluters to bear the costs of abating emissions.

- To collect the financial resources needed to subsidize environmental protection at the regional level (such as municipal wastewater treatment facilities, solid waste disposal and cleanup facilities, and environmental monitoring systems).

These two purposes—providing greater incentives and raising revenue—

are somewhat contradictory, and clearly different charge systems are required to achieve each one fully. The principal incompatibility between the two is pointed out in the recommendations of the Organisation for Economic and Co-operative Development (OECD) on applications of economic instruments for environmental policy: "It should be underlined that when environmental charges do reach incentive levels (i.e., achieve environmental objectives), earmarking revenue would lead to economically inefficient solutions by inducing over-investments and over-expenditure in pollution control." The Russian charge system embodies a strange compromise between these two purposes, with rather unclear priorities given to each. The compromise is especially strange given that another purpose of incentive charges is to improve environmental quality by redistributing revenues. Multiple goals can lead to conflicting design and implementation activities.

In any event, the revenues going to the Russian funds are too small to cope with Russia's severe environmental problems. Following the 500 percent increase in rates, the government issued a decree revising the charge system beginning in 1993. Under the revised system, the charges for emissions within the MPD level are included in the costs of production, whereas under the previous system, all charges were collected from profits.

The revisions in the pollution charge system are obviously intended to regulate how the costs of pollution control are distributed between producers and consumers during the transition period. The price elasticity of supply and demand under current market conditions is very low, and even partial inclusion of pollution charges in production costs will almost automatically lead to a corresponding increase in product prices and thereby transfer a share of the costs of pollution control from producers to consumers. The fact that certain producers have virtual monopolies in energy, petrochemical and other polluting industries facilitates such transfers.

Whether this regulation will succeed in establishing in Russia the "polluter pays" principle depends on factors that are hard to predict. Probably the most important one is the ability to enforce the environmental laws and regulations and the efficiency of general tax policy. However, if the Russian public refuses to allow prices to increase to account for the costs of pollution control, then neither command-and-control nor discharge fees can work.

In sum, the revisions the government made to the charge program in 1992 were all intended to adjust it to the economic situation in Russia. As far as the increases in the charge rates go, observers in the ministry suggest the rates should be indexed to inflation. It is unclear whether these changes will work in the long run, given the many economic problems in Russia that dwarf the pollution control problems and the discharge fee reform.

Costs of Program Implementation

The Russian pollution charge initiative is not set up financially and institutionally within the MEP&NR as a separate special program. Expenditures on implementation are lumped in with the estimates of expenditures for MEP&NR's overall operations (central and territorial offices) and for the research institutions of the ministry and those of other entities such as the Russian Academy of Sciences and the Ministries of Science, Health Care and, Agriculture. Some projects related to the pollution charge program are funded through agreements between a research organization and a sponsor to accomplish a specific task (for example, determination of the coefficients used to correct the charge rates for a specific type of pollutant or specific area).

In short, it is very difficult to estimate and break down the total costs associated with the development and introduction of the charge system, because in the federal budget, they are part of the aggregate expenditures to maintain the MEP&NR and to fund federal and regional research and design programs. According to the forecast for 1993, expenditures for the MEP&NR are 6.1 billion rubles, and for research and design 10.6 billion rubles. Russian experts assume the share of direct and indirect expenditures to develop, implement, and maintain the pollution charge system is not more than 5-7 percent of these sums. Thus, the pollution charge program costs for 1993 will be nearly 1 billion rubles, or about $800,000.[14]

Summary of the Administrative Problems

The EPL and pollution charge regulations were researched, introduced on an experimental basis, and promulgated during the transition from a command-and-control to a market economy. The program was in fact designed for a transition economy. Its authors realized the program must generate its own capital, create incentives for enforcement at every level, target polluters, and create incentives for changes in behavior.

Given Russia's economic difficulties, the government decided not to make the financial goal of the pollution charge program large enough to permit all the country's environmental problems to be solved. Instead, the goal was to generate enough revenues to finance critical projects, such as the construction of water treatment facilities and cleanup of hazardous waste sites. Within this context, the charge system has worked to the satisfaction of national and local authorities, except in the area of adjustment for inflation.

Developing a domestic environmental goods and services industry is of continuing importance to the MEP&NR. For an incentive-based pollution charge system to be successful, companies must be able to avoid paying the charge by implementing new practices, technologies, and equipment to control their pollution. In Russia, only a handful of private environmental firms have emerged to satisfy the still small demand for remediation services. Russia has nothing like the array of options for effectively reducing the costs of pollution

---

[14] By way of context, the draft budget for the Russian Federation for 1993 provides 465.5 billion rubles to maintain the entire government structure, including 90 billion rubles for tax collection.

that industries have in the West, where the environmental market is measured in billions of dollars.

Primarily for this reason, the MEP&NR has placed a high priority on developing a private environmental sector. Another reason is enforcement. The ministry hopes competing measurement technologies will emerge and that standards can be established soon, so that the burden of proving violations is eased. Without cost-effective measurement technologies, firms will be able to avoid detection and subsequent enforcement actions. At present, Russia has no generally accepted technological standards for measuring emissions, effluents, and solid waste. Not only is equipment important, but so are good laboratory procedures, standardized testing protocols, procedures for logging and analyzing data, and assessment techniques.

To summarize, the main administrative problems with the charge program are:

- The lack of an appropriate system—equipment, methods, and personnel—for monitoring discharges.

- Inadequate equipment and expertise of inspection personnel, a situation that makes the identification and punishment of environmental violators difficult.

- An inability to enforce the collection of charges because of uncertainty and contradictions in the legislation. Even where the amounts of emissions have been precisely determined, enterprises do not always pay the charges. One reason is that the regulations leave a great deal of room for "compromise" among the enterprises, banks, and inspection services. The other reason is the legal impossibility of forcing an insolvent enterprise to pay charges (and other taxes), even if the law requires it. As a result, only 3.0 billion rubles in revenues accrued to the environmental funds in 1992, less than half the expected amount.

- The absence of a clear distribution of authority between the federal and territorial levels. Although legally the central ministry runs the program, establishes the base rates for charges, and approves the procedures for collecting the fees and allocating the revenues, the main decisions are made at the regional and local levels. Moscow has chosen as its role to provide moral and methodological support. Regional and local environmental regulators and fund directors say they must have wide legal rights to set and adjust charge rates and manage environmental fund revenues.

- The absence of clear regulations spelling out how to distribute the environmental protection costs among polluters, the federal and regional budgets, and the federal and regional environmental funds. As such, it is difficult for the environmental funds to allocate their revenue properly. In turn, enterprise managers try to get more subsidies from the environmental funds.

- Unresolved questions regarding economic liability for the environmental damage resulting from an enterprise's previous and current technology. These questions will become more crucial as privatization advances.

- Insufficient institutional support. The ministry does not have either a special staff training program or a special implementation program. It has not studied the experience and problems with implementation systematically, it has only a rudimentary oversight system, and it has not compiled the administrative costs.

- Excessively complicated charge system, in part because of the inclusion of hundreds of types of pollutants. The charge rates are calculated with a precision beyond the capacity of monitoring and inspection.

- Erosion of the pollution charges by inflation. The 500 percent increase in charge rates in 1992 was insufficient to offset inflation. The pollution charge is not as well-defended against inflation as are the high-priority taxes, such as the value-added or profit taxes. The rates for these taxes are set as a percentage of producers' income. As a result, revenues automatically increase at the same rate as inflation.

PROGRAM RESULTS

The pollution charge system in Russia is designed to improve environmental outcomes in the areas of air, water, and solid waste disposal. Given the brief time the program has been in place and poor quality of the data—ambient and source—on prior environmental conditions, it is difficult to measure the program's success in terms of environmental outcomes. Further, since a variety of important benchmarks for administrative, environmental, financial, and institutional circumstances are absent, any evaluation must be anecdotal. So far, no nation that has instituted a pollution charge program can quantify the direct effect it has had on pollution abatement. Such evidence is even less likely in a country where the pollution charge program is being introduced during an economic crisis.

With these caveats, the consensus among central and local environmental officials and experts in Russia is that the program, despite its serious problems, is probably the most effective instrument the MEP&NR has. According to various sources, the program has had the following positive results:

(1) Those interviewed and regulators believe that because the government implemented the charge program at the same time as the economic reforms aimed at increasing the legal liability of enterprises, directors and financial managers are paying more attention to pollution control.

(2) The charge system is clearly a good way to raise revenues earmarked

for environmental purposes. Government has cut back on expenditures for environmental protection under the federal and regional budgets. As a result, financial support from the non-budgetary funds has become extremely important for many enterprises.

(3) Whereas previously government virtually ignored environmental standards and quotas in decision-making, the pollution charges have made them more important.

(4) There is some evidence the fees are high enough to influence the behavior of polluters, although the exact effect probably varies widely across regions.

Currently, the program's failures and problems far outnumber its achievements. Under these circumstances, the MEP&NR's optimism seems ill-placed. However, when the outcomes of the charge program are compared with the limited success of other tax innovations and environmental regulatory efforts, the MEP&NR's position is more understandable. Further, MEP&NR staff and most other experts interviewed judge the program not by past or current experience but in terms of future potential. However, it is very difficult to predict how effective the charge program will be.

It is important that Russia come to realize that pollution charges and other economic instruments for environmental management have no future unless two important conditions are present:

(1) Institutions and people must be made legally and economically liable for the environmental consequences of their business activities, with stringent regulations that distribute this liability across government agencies, private institutions and foreign investors.

(2) The environmental consequences of business activities must be monitored and regulations enforced.

THE NEED FOR PROGRAM EVALUATION

According to the Russians who participated in the formulation of the pollution charge system, evaluation is rarely practiced in Russia. Most reports on program performance look solely at resource expenditures and numbers served. Seldom do public agencies provide timely information on the quality and outcomes of programs. In the absence of attention to quality and outcomes, programs too often become wasteful, ineffective, and unresponsive, and the credibility of government suffers.

The Russian pollution charge program is unique, as are Russia's economic, environmental, legal, and social conditions. Common evaluation criteria may not apply at this time because information about the program is lacking. Nevertheless, evaluation at some level is extremely important.

A way to begin to evaluate the program is with two sets of questions:

(1) Do pollution charges and corresponding earmarked environmental funds promote actions to reduce environmental problems under current conditions?

(2) Will the program continue, and if so, how should it be modified?

Answering these two questions requires the design of an evaluation system. Such a system would have a number of benefits, such as:

- Documenting the goals, institutional characteristics, and performance of the program.
- Guiding decisions on the allocation of resources.
- Providing a basis for program modifications to increase effectiveness and efficiency.
- Providing information about the relationship of a market incentive program to other regulatory strategies.
- Examining the import of political, economic, and social structures in a time of transition on the applicability, efficiency, and effectiveness of economic incentives.
- Testing the theories underlying the program.
- Serving as a medium of accountability.
- Developing an evaluation research methodology and system that can be transferred to other market-based programs or used in conjunction with other regulatory schemes.
- Acquiring hands-on experience through the hands-on assistance provided by an independent evaluation team.

An important outcome of evaluation would be the development of a cadre of Russian evaluators trained in the processes, procedures, and techniques of evaluation. The Russian evaluators would then teach others to focus on results through monitoring and measurement of programs relative to targets. The Russians believe the time is right for training trainers, who will then introduce others to various aspects of program evaluation.

Another significant outcome would be an analytical description of the fee program that could be shared with policy-makers from the United States and other countries. For example, answering questions relating to the program's context, origins and history, rationale, goals and objectives, personnel, and budgetary and administrative arrangements will provide empirical validation of this theoretical approach to regulation. Given the degree of interest in the program, the results of the evaluation are likely to be welcomed by policy-makers in many nations. Too often, effective governmental actions are hampered by incomplete information, uncertainty, weak institutions, and inadequate administrative capacity. A third outcome would be written reflections on how political, social, economic, and industrial structures such as ownership and competition have affected the effectiveness of the charge system. For example, how does the subsidization of heavy-polluting sectors of the economy affect the proposed charge system?

Assuming both that the above conditions are satisfied and that information is available, it would be helpful for experts to review the Russian system of direct regulations and economic instruments for environmental protection. The purposes of the review would be to (1) discover whether the program is consistent with economic theory; (2) determine whether the charges are high enough to affect the behavior of firms or consumers; (3) determine whether the redistribution of revenues improves environmental quality; and (4) find out whether polluters have instituted less expensive processes to meet environmental standards than they might have under a system in which regulations and standards play a dominant role.

CONCLUSIONS

The pollution charge system is unique for many reasons, not the least of which are the twin difficulties of implementing a system at a time of severe economic and social upheaval without some of the key building blocks of a sound environmental program. Whether Russia chooses to base its environmental program on command-and-control or pollution charges, it cannot wait to tackle environmental cleanup until economic, social, administrative, and technological circumstances change. The job needed to start yesterday, and it must be worked on today, even with imperfect tools. This beginning will be a start in the right direction.

The pollution charge system has become the cornerstone of environmental protection programs in Russia. As such, it will no doubt grow in sophistication as new social and economic realities take form. Already the system is revealing interesting results, but only by constant oversight can these results provide meaningful information to Russian decision-makers and interested observers.

It is very important that Russia adopt routine oversight systems that capture program outcomes in real time and provide periodic updates on the program's successes and failures. Given the increasingly large audience for information on the use of pollution charges, such an investment can yield big dividends to both Russia and all environmentally threatened countries.

APPENDIX. RUSSIAN SOURCES

This section on the Russian pollution charge system includes information from interviews and discussions with the following people:

A. Averchenkov, Deputy Minister, Ministry of Environmental Protection and Natural Resources

Ye. Groshev, Official, Ministry of Environmental Protection and Natural Resources

O. Kukushkin, Official, Ministry of Environmental Protection and Natural Resources

A. Shevchuk, Official, Ministry of Environmental Protection and Natural Resources

B. Itkin, Official, Ministry of Environmental Protection and Natural Resources

A. Ignatiev, Director of the Institute for Natural Resource and Environmental Economics, Ministry of Environmental Protection and Natural Resources

Ye. Shopkoev, Chief of the Environmental and Natural Resources Department, Russian Federation Ministry of Economy

A. Salina, Chief of Department, Russian Federation Ministry of Finance

S. Vasiliev, Deputy Chairman, Moscow Committee for Environmental Protection and Natural Resources

V. Pozdnyakov, Deputy Chairman, Tver Committee for Environmental Protection and Natural Resources

L. Fomenko, Deputy Chairman, Yarslavl Committee for Environmental Protection and Natural Resources

I. Semenikhin, General Director, Federal Environmental Fund

V. Tretiakov, Chief of Department, Russian Federation Committee on Land Resources

O. Balatzkiy, Research Director, Technological Institute of Sumy, Ukraine

A. Bobylev, Professor, Moscow State University

A. Golub, Director, Center for Environmental Economics Studies

3
CHARGES FOR SERVICES AND DEPOSIT-REFUND SYSTEMS: TWO CASE STUDIES

In theory, economic incentives can be applied to affect the pricing and, therefore, the use of resources so as to have a positive impact on the environment. In practice, however, a large gap may exist between theory and reality. For example, charges for services and deposit-refund systems in practice may act not so much as monetary incentives as psychological ones.

The focus of this chapter is two economic incentive programs being used to control solid waste. King County, Washington, has implemented a variable fee for the collection and disposal of solid waste. The state of Michigan is using a deposit-refund system for beverage containers. The basis for the variable fee imposed by King County is the number of trash cans a household puts out—a household pays a certain amount for the pick-up and disposal of one can of trash and a considerably higher amount for a second can of trash. King County instituted this system to reduce waste and encourage recycling. King County also uses two other instruments that can be classified as economic incentives. The first is a "tipping fee," or charge, that is assessed for unloading trash at a disposal, incineration, or transfer site. Tipping fees have risen fourfold since the 1980s. The second is the creation of markets for commodities such as green glass and mixed wastepaper.

Michigan's bottle bill requires a 10-cent deposit on beverage containers, including beer and soft drink containers. The industries that produce and sell the product and the consumers fund the system. In a deposit-refund system, purchasers of potentially polluting products such as beverage containers make a deposit on products when they buy them. When they or others return the containers, they receive a refund of the deposit. The aim of the system is to encourage consumers and others to return containers and thereby cut down on this source of pollution. Refund of the deposit acts as a reward for people who practice environmentally correct behavior. Interestingly, the majority of states with forced deposit laws have enacted them to lessen litter and reduce the volume of solid waste, although the original rationale for deposit-refund systems for beverage containers was financial—returnable and reused bottles are less expensive than throwaways. Deposit-refund systems for products such as batteries and residuals from pesticide containers have also been used to prevent the release of toxic materials into the environment.

It is difficult to sort out the effects of King County's variable fees on the recycling of materials. One reason is that the variable fees are just one small element in a large regime of tools for managing solid waste that includes education, public awareness, and alternative ways to handle trash. The jury is still out on whether higher rates for additional trash cans have encouraged recycling. How the rates fit into the county's comprehensive solid waste management program has not been determined, either. Similarly, it is difficult to determine whether Michigan's 10-cent deposit per container has affected the rate of return. In other words, whether the fee levels or the deposit-refund amounts are acting as incentives for better waste management, or whether other non-monetary incentives such as psychological or political motivations are having a greater effect on the performance of these instruments, is of considerable interest.

Both case studies point out the need to assess carefully the relationships between economic instruments and other forms of regulation. For example, advocates of comprehensive solid waste recycling programs contend that deposit laws negatively affect recyclable waste programs because they draw off beverage containers, which have a much higher scrap value and much steadier markets than other recyclable products do. Further, they contend that deposit laws do not reduce total litter and solid waste inasmuch as beverage containers constitute only a small part of roadside litter, which includes food packing containers, newspapers, cigarette packages, lids, straws and so on. In that regard, programs such as "Adopt a Highway" and other outreach efforts have been more successful. In short, supporters of comprehensive recycling programs argue that deposit programs and comprehensive recycling efforts do not mix.

Another issue both case studies raise is their effectiveness in generating revenue for environmental management. Are the funds generated earmarked for environmental protection? Are they linked to amelioration of the type of negative effects the charge is supposed to mitigate? This latter issue is tied to an argument regarding the openness and transparency of instruments and the fear that the revenues generated will become another source of general revenue or will create a windfall for a particular target group. King County's Solid Waste Division has profited from recycling. In Michigan, the bottlers and distributors have benefitted from the unclaimed deposits. In both cases, other groups have lost. In King County, the small recycling firms that prospered in the 1970s have gone out of business because they were not able to afford the capital equipment necessary to become a player in the larger curbside recycling efforts. Michigan's retailers have had to bear the costs associated with handling and storing returned containers.

The two case studies that follow identify and explore these and several other issues, including the involvement of industry at different stages of the incentive system, the availability of information about the objectives of the system, the impact of information on the acceptability of the system, manpower requirements, and the need for evaluation. The information should be of use to policy-makers and regulators looking to apply these instruments to their particular localities. Although the case studies are intended to help the reader answer the question of whether the programs achieved their intended outcomes, results are not the only feature of the program that merits attention. In fact, what these cases best bring to light are the strengths and weaknesses of their implementation.

CASE STUDY:
ECONOMIC INCENTIVES FOR MANAGING SOLID WASTE IN KING COUNTY, WASHINGTON

Every day, Americans throw away an average of more than three and a half pounds of trash each, a pound more than Germans do and almost a pound more than Americans generated in 1960. About 80 percent of this mountain of garbage goes to landfills. The landfills are, however, filling up rapidly, at the same time that new environmental standards make it more expensive to build and operate them. The "not-in-my-backyard" stance makes it more difficult to find new locations.

For the last several years, local governments, which have responsibility for managing solid waste, have been looking for alternatives to landfills, and the states and federal government have been encouraging efforts to reduce waste and recycle as much as possible. In the 1980s, the charges for using landfills (or tipping fees) rose sharply, and in more than 200 cities, homes now have to pay more for pick-ups of two cans of trash than of one. The states of Washington, New York, and California, along with some local governments, have been actively seeking to create markets for recyclable waste.

Collectively, local governments have accumulated extensive experience with economic incentives for controlling solid waste, perhaps more so than in any other environmental area. King County, Washington, and Seattle, the state's largest city, are among the first large local governments to rely on economic incentives as a key tool for solid waste management. This section looks at the experience of King County in implementing economic incentives for solid waste management. The next section describes the county's experience. It begins with a brief summary of how King County and Seattle came to emphasize recycling and use economic incentives to encourage residential customers to separate recyclable waste at the curbside. It also describes the physical system for handling solid waste in the county; the roles of the municipalities, county and state; and the economic incentives used, including higher tipping fees and variable rates based on the number of trash cans picked up.

The third section analyzes the impact of these economic incentives on the staffing and management of the responsible agencies; changes in the mission and roles of the agencies; and changes in the distribution of risks, costs, and "profits." So far, the county and towns have not run into major difficulties in instituting the economic incentives. The implementing agencies, however, have undergone many changes as a result of their commitment to recycling.

The fourth section describes the need to evaluate the measures the county and cities have taken to reduce waste and promote recycling. Although King County is gathering far more information than ever before about the waste stream and the costs of handling it, there is still little information about the costs and effectiveness of its waste reduction and recycling programs. This section proposes an evaluation strategy that would help counties and cities choose the most effective way to meet their ambitious recycling and waste management goals.

The conclusion suggests lessons that might be drawn from King County's experience. It also raises two questions:

- Are the incentives really economic, or are they more accurately described as psychological?
- Do the incentives seem to be leading toward a more efficient use of society's resources, as the theory of market incentives suggests would happen?

THE MANAGEMENT OF SOLID WASTE IN KING COUNTY

For decades, King County, like many other local governments, dumped its trash into landfills that were subject to minimal environmental controls. This practice came to an end in the early 1980s, as, one by one, the landfills in western Washington either closed or faced the need to spend millions on environmental controls. Businesses in Seattle had been sending their waste to a privately owned landfill north of King County until 1978, when it shut down. (It was later designated a Superfund site.) Residential waste from Seattle then went to a facility the city operated in a nearby suburb until 1986, when it closed.

Most of the waste from both these landfills was diverted to Cedar Hills, a large landfill operated by King County. Because Seattle accounts for about a third of the county's population and solid waste, this move shortened the likely future life of Cedar Hills by several years. In addition, the county had its own environmental problems. Between 1978 and the early 1980s, environmental groups and homeowners living near Cedar Hills filed several lawsuits about traffic and environmental problems. There were also problems at the smaller landfills the county operated in rural areas. In 1985, the state of Washington adopted "minimum functional standards" that forced such measures as control of drainage, venting of the gases that build up inside landfills, and a plastic liner and two feet of clay under a landfill and, in the case of closed landfills, on top as well.[1]

---

[1] The state was far ahead of the EPA, which did not promulgate comparably tough standards until 1991.

Rod Hansen, who became manager of the King County Solid Waste Division in 1983, pressed the county to adopt a policy that met and even exceeded existing and possible future environmental requirements. As a result, the county found itself having to raise much more revenue to finance environmental improvements. Eventually the county spent more than $95 million to clean up Cedar Hills and another $29 million for remedial upgrades at four rural landfills.

As it became more difficult and expensive to operate the landfills, and as it became clear that neighborhoods might organize to prevent the construction of a new landfill when Cedar Hills filled up, both King County and Seattle began to rethink their solid waste policies. The classic environmental approach—characterized by both a top-down, command-and-control scheme and a bottom-up, not-in-my-backyard stance—forced Seattle and King County to look for new ways to manage their trash.

Initially, both decided to build incinerators to burn trash. Ever since the energy crisis of the early 1970s, King County had been working on plans for "waste-to-energy" plants that would generate electricity by burning garbage. The new county executive elected in 1986 directed the King County Solid Waste Division to stop studying and start planning in earnest. The county released a preliminary proposal to build as many as four incinerators in different parts of the county, and Seattle issued a proposal to build two of its own.

All hell broke loose. When solid waste officials identified a half-dozen possible sites, they aroused a half-dozen not-in-my-backyard groups that quickly organized themselves into the Alliance for Solid Waste Alternatives, a county-wide federation of neighborhood groups and environmentalists that did not want incinerators anywhere. This federation worked with an older environmental group, Washington Citizens for Recycling (which still has a busy office of a half-dozen paid staff and volunteers). The protesters attended hearings en masse and picketed meetings of the county council. Solid waste management became the central issue in local politics, and an important state issue as well.

Bowing to public will in January 1987, local officials reversed directions. Charles Royer, the progressive new mayor of Seattle, and Paul Barden, a conservative "law-and-order" county council member from the suburbs south of the city, led the change. The county organized citizen task forces on all aspects of solid waste management. By the end of 1988, the city and the county each made a formal commitment to waste reduction and recycling as the first two choices for managing trash, with incinerators and landfills following.

City and county officials began to plan and institute new recycling programs. The state Legislature also went to work, framing legislation that in 1989 became the Waste Not Washington Act. This act endorsed the same hierarchy of preferred options for managing solid waste and called on all counties to work with municipalities to prepare comprehensive solid waste management plans consistent with these priorities. It also set as a statewide goal for 1995 that 50 percent of all solid waste should be recycled. Seattle set a more ambitious goal of eliminating, recycling, or composting 60 percent of its total waste stream by

118   *The Environment Goes to Market*

1998. King County's goal is to recycle or eliminate 65 percent of the waste stream by 2000.

How Solid Wastes Are Managed

The physical arrangements for handling solid waste in King County are complex, and the legal authority to regulate the system is fragmented. With respect to the *collection* of trash, except for the small rural town of Enumclaw, it is handled by private firms, not government agencies. Incorporated cities can decide to contract with firms to collect trash, set the rates and conditions of collection, and require that homeowners pay for pick-up, including separate pick-up for recyclable materials. In unincorporated areas and where cities do not exercise their authority to contract, the state issues franchises to private firms giving them the exclusive right to collect garbage in specific areas. The Washington Utilities and Transportation Commission (WUTC), the state public utility commission, sets the rates and terms of service.

Although a city may require a homeowner or business to pay for garbage pick-up, it does not require use of the service for all trash. Anyone can sell or donate their trash or can take it to a transfer station or landfill. Many large businesses contract directly for the collection of their trash. Some individuals and businesses sell their trash to recycling firms and scrap yards. Safeway, for example, holds auctions every few weeks to sell the used cardboard boxes from its many stores; the auctions generate thousands of dollars each month. Boeing Co., the aircraft manufacturer, has set up a small staff to assemble monthly packages of recyclable products, which it offers to the highest bidder. A vigorous small-scale recycling industry emerged in the 1970s that collects high-value products such as aluminum cans and newspapers and sells them to manufacturers, producers of cans, or pulp mills. Many churches and scout troops raise funds by collecting these materials.

In short, the collection of trash is largely the responsibility of cities, private firms, and the WUTC. The county has no direct role in collecting trash other than preparing a county-wide solid waste management plan.

The county does, however, have a central role in the *disposal* of trash. County ordinances require that, except for construction debris and materials for recycling, commercial trash must go to Cedar Hills. Further, most residential trash, except from Seattle, also goes to Cedar Hills. The trash flows through its seven county-run transfer stations or rural "drop boxes." From those points it is trucked by county employees to the Cedar Hills landfill. (Some residential and commercial trash is trucked directly to the landfill by private collectors.[2] Homeowners are not allowed into Cedar Hills; they must go to the transfer stations.) The county levies tipping fees for these disposal services. In urban areas, the fees are included in the bills that cities charge their residents for garbage collection. In rural areas, the haulers bill residents directly.

---

[2] It is possible that haulers may illegally divert small amounts of commercial waste to privately owned landfills.

Until 1986, the cities sent their trash to Cedar Hills under an informal arrangement. As the volume of trash rose and major investments in environmental controls became necessary, the county decided it needed to be able to make firm projections of the volume of trash and life of the landfill. It therefore entered into an agreement with the suburban cities whereby they will send all their trash to the county landfill (or another site chosen by the county) for 40 years. Seattle signed for only five years, until 1991. Since then, Seattle has been contracting with Waste Management, Inc., to transport all its non-recycled residential waste to a WMI landfill in Oregon. (WMI is now planning to develop a landfill in eastern Washington to avoid exporting the trash.)

Incentives for Waste Reduction and Recycling

King County uses many different devices to reduce the flow of trash into Cedar Hills. Perhaps the best-known is a separate pick-up for recyclable products. Curbside recycling was pioneered elsewhere, but Seattle was the first large city to use it, followed quickly by the suburbs. King County's solid waste management plan is now built around comprehensive recycling. This plan has made it possible for the suburbs to adopt various curbside recycling programs. Some require that homeowners put glass, paper, tin cans, and other materials into separate containers, while others require that recyclable products be separated en masse from the rest of the trash. Some forbid mixing recyclables with trash, and still others make separation voluntary.

In addition to requiring some kind of curbside recycling, King County is using three measures that can be classified as market, or economic, incentives. The first is higher tipping fees for discarding trash. These fees have risen about fourfold since the early 1980s (table 3-1). The impetus for raising the fees was not so much to encourage recycling, however, as it was to cover the costs of cleaning up the landfills and building a reserve fund for incinerators. The higher tipping fees show up in higher bills for garbage collection.

The second incentive is a variable rate for collection based on the number of trash cans picked up. Since haulers do not bill separately to pick up recyclables, a higher fee for a second trash bin encourages people to go to the trouble of separating recyclables from the rest of their solid waste. Some private trash collectors have long charged a little more for two trash cans instead of one, and the fee schedule adopted by Seattle rises steeply. King County has strongly encouraged the suburbs to impose variable trash can rates in their contracts with trash collectors. The standard recommended in county policy is for rates to be 40 percent higher for a two-can pick-up than for a one-can. County officials encourage cities to use a 100 percent mark-up. (Table 3-2 shows the rates charged for residential trash collection.)

The third incentive consists of an array of activities that are directed not at the supply of recyclable materials but at the demand, that is, at what happens to recyclables after haulers pick them up. When Seattle and King County adopted recycling and waste reduction as their preferred approach to

TABLE 3-1
KING COUNTY SOLID WASTE DISPOSAL FEES, 1981-92
(dollars)

	Jan. 1981	Jan. 1982	Jan. 1983	Dec. 1986	June 1990	Jan. 1992
Sites with fee scales (per ton)						
Basic fee—solid waste	15.00	18.50	26.50	47.00		66.00
Yard waste separated at source[a]					31.00	58.00
Fee for charitable organizations[b]						43.00
Sites without fee scales (per cubic yard)						
Compacted solid waste	4.50	5.60	7.90	14.00		19.00
Uncompacted solid waste	2.50	3.10	4.40	8.00		11.00
Compacted separated yard waste[a]					9.60	17.00
Uncompacted separated yard waste[b]					5.25	9.50
Minimum charges (per entry)						
Solid waste	2.00	2.50	3.50	6.50		9.28
Yard waste separated at source[a]					4.00	7.41
Charitable organizations[b]						5.93
Cedar Hills charges						
Direct from region (per ton)	5.50	7.00	11.00	31.50		43.00
Other vehicles (per ton)	15.00	18.50	26.50	47.00		66.00
Minimum charge (per entry)	2.00	2.50	3.50	6.50		9.28
Special waste (per ton)					75.00	100.00
Special waste minimum charge (per entry)					10.50	13.86

[a]There was no separate fee for yard waste until January 1991.
[b]There was no separate fee for charitable organizations until January 1992.
Source: King County Department of Public Works, Dvision of Solid Waste, Washington.

TABLE 3-2
RESIDENTIAL SOLID WASTE
AND RECYCLING COLLECTION RATES
(dollars per container)

Jurisdiction	Mini-can	1 can	2 cans
Seattle	11.50	14.98	29.96
Suburbs			
Algona		7.05	9.70
Auburn	6.50	7.00	15.80
Beaux Arts	8.50	9.70	11.95
Bellevue		11.75	16.15
Black Diamond		8.10	10.15
Bothell		10.00	14.00
Carnation		11.15	15.00
Clyde Hill	8.89	10.43	14.22
Des Moines		7.10	9.85
Duvall		7.62	8.90
Enumclaw (2-can minimum)			10.05
Federal Way		7.10	9.85
Hunts Point	5.00	7.85	10.85
Issaquah	7.92	12.78	22.51
Kent		7.60	11.35
Kirkland	6.35	10.80	15.20
Lake Forest Park	6.35	9.95	13.95
Medina	5.00	7.85	10.85
Mercer Island	6.35	10.80	15.20
Milton		6.15	9.34
Normandy Park	5.60	7.40	11.10
North Blend		10.00	
Pacific	5.60	6.95	10.95
Redmond	7.14	11.55	16.80
Renton	3.60	8.90	14.90
SeaTac	5.60	8.35	11.75
Skykomish		9.50	
Snoquaimie		10.35	
Tukwila	7.10	10.65	14.20
Yarrow Point	5.00	7.85	10.95
Unincorporated King County			
Service Area 1	8.21	12.21	16.21
Service Area 2	5.22	8.07	11.07
Service Area 3	7.21	10.36	14.26
Service Area 4	8.20	12.93	17.18
Service Area 5	7.64	11.54	15.29
Service Area 6	8.27	12.32	16.87
Service Area 7	7.32	10.32	14.42
Service Area 8	6.05	9.60	13.35

Note: Blank spaces mean there is no separate rate.

Source: Draft 1992 Comprehensive Solid Waste Management Plan and EIS, King County Department of Public Works, Solid Waste Division, vol. 1, August 1992, pp. IV-5.

solid waste management, citizens responded eagerly. As a result, the markets for most recyclables became glutted, and the prices for some materials, especially green glass and mixed wastepaper, fell sharply. In response, both the county and state have undertaken to build more markets for these elements of the waste stream. King County organized the King County Commission for Marketing Recyclable Materials, and the state established the Clean Washington Center as a division in the state Department of Trade and Economic Development.

In addition to the regulatory requirements, curbside recycling programs, and economic incentives, the county and cities have invested heavily in public education. Many of their efforts are broadly targeted at the populace as a whole. Others have a narrower audience. One of the most popular and reportedly the most effective is aimed at children in public schools. (Box 3-1 lists the public education programs.)

BOX 3-1
EDUCATIONAL AND RESEARCH PROGRAMS
TO ENCOURAGE WASTE REDUCTION AND RECYCLING

- **Home Waste Guide**—A widely distributed booklet for homeowners on waste reduction and recycling options and contacts for further information.
- **Recycle Week**—A program of achievement awards for companies, schools, and other institutions for outstanding contributions to waste reduction.
- **School Programs**—Curricula for kindergarten through grade 12.
- **Technical Assistance Program**—Consultations and written materials for businesses.
- **County Model In-House Program**—Measures taken at county agencies to set an example of ways to recycle and reduce waste.
- **Master Recycler/Composter**—Training manual for volunteers.
- **Composing Bins**—Free bins and information on how to compost household garbage.
- **Cloth Diaper Project**—Workshops and provision of cloth diapers to low-income families.
- **Dollars for Data**—Financial aid to businesses for implementing and measuring the results of waste reduction projects.

Source: *Draft 1992 Comprehensive Solid Waste Management Plan and EIS*, King County Department of Public Works, Solid Waste Division, vol. 1, August 1992, pp. III, 2-3.

Although the Waste Not Washington Act endorsed waste reduction along with recycling, it removed one of the key tools local governments could use. The statute includes a "ban on bans" and a "ban on fees" (for example, bottle deposits) that eliminate local authority to regulate the sale of products that governments might want to keep out of the waste stream. Without the ability to impose bans or fees, local governments have had few means to force waste reduction. The provision sunsets in 1993, and the local governments are lobbying hard to see it disappear.

At the local level, county council member Barden sought an ordinance to ban all paper diapers from the Cedar Hills landfill but could not get the bill through. However, the county Department of Health was successful in banning *dirty* paper diapers from the landfill on public health grounds, and one city kept a fast-food franchise from using plastic foam cups and packaging by making the ban a condition of a zoning permit.

ADMINISTRATIVE AND POLITICAL EFFECTS OF THE NEW POLICIES

After the battles of 1987 quieted down, the shift to recycling and waste reduction moved forward without much organizational strain or political conflict. Implementation of the policy of waste reduction and recycling does not appear to have been especially difficult. County and city officials say they are proud of their progress. For many public employees, the new directions seem to have provided opportunities for personal and professional growth. Several of the citizens who led the fight against incinerators are now actively involved in various advisory groups and praise the division openly. As one said, "We have some quibbles with what some of the suburban cities have done, like still charging a flat fee for trash collection, but basically we think King County is doing just fine."

At the same time, the new policies have required major changes in most of the agencies in King County and suburban cities that have implemented them. They are hiring new kinds of staff, reorienting their management objectives and organizational culture, entering into new kinds of intergovernmental and public-private relationships, and encountering new kinds of risk. Some of these changes are driven directly by the use of economic incentives, but most are the result of the shift in policy generally.

Changes in Staffing

In the past eight years, the budget of the King County Solid Waste Division almost tripled and staff more than doubled. The increase in revenues is the result of the higher tipping fees. The increase in staffing has occurred at two levels. First, the field staff has been expanded (table 3-3), primarily by adding people to manage the environmental control programs. Second, the headquarters staff has grown rapidly: in 1991 the number of full-time employees was around four times larger than it was in 1985 and eight times what it was in 1983. It includes many new environmental engineers (up from 2 in the mid-1980s to 20 in 1993), people working on waste reduction and recycling programs, and a larger information management unit.

TABLE 3-3
KING COUNTY DIVISION OF SOLID WASTE

	Revenue ($)	Expenditures ($)	Admin. Staff[a]	Field Staff[a]	Tons of Solid Waste
1985	19,914,530	17,937,358	23	154	887,591
1986	25,543,367	24,322,600	25	156	958,130
1987	54,284,413	51,921,332	34	208	1,341,674
1988	54,097,657	52,481,457	38	229	1,302,507
1989	55,873,548	52,882,933	51	230	1,285,834
1990	63,774,532	65,159,417	76	264	1,433,522
1991	61,737,545	65,794,498	91	287	1,186,600

[a]Full-time equivalent.
Source: King County Solid Waste Division, Department of Public Works.

The cities have also hired additional staff to handle their recycling programs. Before 1987, most city councils spent little time on solid waste issues, other than periodically discussing new contracts for collection services. Eventually, most cities had to hire full-time staff to coordinate the recycling. Initially, they financed this staff mostly through grants from the county or the state Department of Ecology. Subsequently most increased trash collection fees to cover the costs. The county helps the cities with smaller grants for program activities.

At first, the role of city recycling coordinators was to work with private collectors to design and implement new recycling programs. For example, the city usually has to choose what kind of recycling bin to use, and there are complex design issues in recycling programs for multi-family homes, such as modifying building codes to ensure enough room for the recycling bins and trash containers.

Once the basic programs were in place, many of the recycling coordinators turned to finding ways to make them more effective, with some success. For example, some coordinators have worked with the owners of apartment houses to put decals inside kitchen cabinets telling new residents how to recycle. Others have helped organize composting programs. Some coordinators have been promoting recycling inside city offices by calling for the procurement of products that contain recycled materials. The county Solid Waste Division has for a long time been holding monthly roundtables at which recycling staff from the county and cities can exchange ideas. Recently, the state Department of Ecology has organized similar sessions.

The other major change in staffing arose as a result of the county and state's effort to develop markets for recycled products. The King County Commission for Marketing Recyclable Materials has a staff of six (its first head was the former director of recycling in Berkeley, California). The state's Clean Washington Center is managed by the former head of the state's small business assistance program and has a staff that includes four engineers.

Changes in Agency Mission and Organizational Culture

According to manager Hansen and several outside observers, there have been dramatic changes in the mission and organizational culture of the King County Solid Waste Division in the past eight years. The most important is that the division is now an environmental management agency. Its role is not simply to provide a public service but rather to manage the solid waste system to meet environmental goals and preserve Cedar Hills as an "environmental asset." County council member Barden expressed a similar view. He said the county has a fiduciary responsibility to the taxpayers for Cedar Hills; it is a valuable asset whose life must be extended.

Along with the change in the division's mission and its addition of staff has come a sharp change in organizational culture. In the words of several of those interviewed, engineers oriented toward "structural solutions" used to dominate the division. When they faced a problem, they thought immediately of building a new facility such as a landfill or incinerator. Now the division's business is to alter the behavior of the public by encouraging and helping it change the way it handles solid waste. According to Hansen, "This has created a need to reeducate the whole agency, not just staff of the recycling program but also the receptionists and the guards at the front gate, so they answer the public's questions in new ways."[3]

Field operations continue to dominate the division. The largest sub-units in terms of staffing and budget are still Cedar Hills, the transfer stations, the trucking fleet, the rural landfills, and the staff of professional engineers who design and manage these facilities (figure 3-1). Most of these sub-units, except for the shops, have more white-collar workers and more women than in the past. The engineering staff of the division has changed as a result of the addition of many people with backgrounds in environmental engineering, water quality, or soils. The work of the engineers has also changed: they spend more time on the frontline with the public, explaining how environmental concerns are being addressed. New staff at the landfill are responsible for managing environmental controls; a half-dozen of them work full-time picking up paper and plastic that blows off vehicles hauling trash into Cedar Hills.

The change in the profile of division employees is especially notable among the downtown staff. It is young (many of the new employees are in their 30s), and many employees have not been there very long (about half less than four years). Hansen has been with the division longer than any other downtown employee.

Many of the new downtown staff have backgrounds in liberal arts or graduate degrees in public administration, planning, or environmental stud-

---

[3] Interview, February 2, 1993.

FTE=full-time equivalent position.
Source: King County Department of Public Works, Solid Waste Division.

ies. Hansen notes that five years ago "there were no professionals in recycling and waste reduction; everyone working on these issues was in the counterculture." Despite the backgrounds of the new staff, the division has not sought to create a new profession for those who are already working for the county or cities on recycling, planning, and waste reduction. Management believes these professionals can operate effectively without prior experience with solid waste issues. The skills necessary to do a good job, says the director of the division's recycling programs, are expertise in project management, good inter-personal abilities that permit collaboration with members of other organizations, and some policy analysis. The division hired an engineer, who soon moved on.[4]

Other changes can be found in the suburban city councils. Before recycling, some saw the councils as having no significant responsibilities for solid waste management. Now these people are players, conscious of their responsibilities and prerogatives. The recycling tasks these city governments have to deal with are not particularly controversial or technical. Terry Lukens, a city council member in Bellevue, one of Seattle's largest suburbs, with a booming downtown

---

[4] The lack of interest in making waste reduction and recycling a profession stands in contrast to fields such as energy conservation. As with recycling, energy conservation employs a variety of techniques to encourage homeowners and businesses to be less wasteful. Recently, a new profession of "demand-side management" has developed within energy conservation that includes state employees, consultants, and experts with electrical utilities who design and operate programs that encourage conservation by homeowners and businesses. As more cities and counties organize for recycling and waste reduction, perhaps a new profession will emerge.

[5] Interview, February 3, 1993.

of high-rises, says, "There is nothing in the recycling program that is anywhere as controversial as affordable housing or locating a new police station."⁵ Nevertheless, Bellevue has established an environmental services commission of citizens who review and recommend to the council steps that should be taken to manage storm water, surface water, and solid waste. The city council also informally promotes recycling. For example, Lukens suggested to the recycling coordinator that the city stop using envelopes with plastic windows, because they are not easily recycled. The new mission and organizational culture deemphasize "structural solutions." However, a noteworthy event within the division is the construction of a new transfer station (box 3-2).

BOX 3-2
THE NEW ENUMCLAW TRANSFER STATION

The county has six old transfer stations. Built in the 1960s, they are not particularly pleasant places to work or have in the neighborhood. They are rudimentary, with sheet metal roofs and sides open to the weather. Trucks and cars dump their trash directly into county trucks parked in trenches below, the bins for recyclable materials are outside, and the gatekeepers work in shacks. Recognizing the need to replace the old transfer stations, the division has invested $11.7 million in its first new facility, at Enumclaw. One aim is to make it so attractive that it will generate little opposition to other new facilities.

The new station is expansive, convenient, efficient, and handsome. It includes a comfortable building for the gatekeeper with a kitchen and bathroom; a brightly colored kiosk where visitors can get information about recycling and waste management; restrooms for visitors in a separate building; well-appointed changing rooms, showers, kitchen, patio and pleasant lunch room for employees; a large covered area for bins for different recyclable materials; and a large enclosed concrete building where visitors can dump their trash onto a floor where it can be bulldozed into a chute to a hydraulic compactor in the basement and then loaded into county trucks. An attractive system of settling ponds avoids contamination of a nearby stream. The building where trash is dumped is designed for the safety and comfort of both visitors and staff. There are several safety features. Hoses are available for customers to clean out the backs of their trucks after dumping their trash.

The top of the building is decorated with a carved frieze. There was a lengthy debate about what to put on the frieze, with suggestions that it include notable sayings of local Native American chiefs or politically correct exhortations (for example, Recycle or Reduce waste). The division settled on something neutral and related to the business: the chemical symbols and names of elements found in trash, such as "AL aluminum" and "SI silicon." The frieze is not a sign of new extravagance on the part of the division. All county construction projects must set aside 1 percent for art. In this case, the funds were used for the kiosk, a sculpture, and the frieze, as well as for attractive signs.

There have been two significant changes in mission and culture at the state level. One was the creation of the Clean Washington Center and its placement in the Department of Trade and Economic Development, rather than in the Department of Ecology. This placement reflects the decision to make recycling an element of economic development as well as a matter of environmental regulation and public works. The Department of Trade and Economic Development includes the usual range of economic development programs, such as finance and technical assistance for small firms, export promotion and finance, and industrial development. The center fits in as one of three programs that focus on specific sectors of the economy (the other two are the tourism division and a new program for the secondary wood products industry). As such, the center represents a unique effort to build a development strategy around an environmental industry.

The second change involves the role of the WUTC. As explained, the WUTC is legally responsible for approving the rates private firms charge to collect garbage in unincorporated areas and in the cities that do not exercise their authority to contract with collectors. As with other public utility commissions, the WUTC was designed to protect the public from abuse by natural monopolies such as private electrical utilities, gas companies, and some trucking and other transportation companies. The WUTC's approach to the fees for garbage collection is to keep them as low as possible while guaranteeing a reasonable rate of return to the private firms. To guard against monopolistic price-gouging, it sets the rates on the basis of the actual cost of providing collection services rather than on the long-term cost or on the environmental effects of dumping trash in a landfill. Because it costs only a little more for a collector to pick up two trash cans instead of one, the WUTC favors only a small mark-up for the second can. This approach conflicts directly with the orientation of the county Solid Waste Division, which prefers sharply graduated rates for trash cans to encourage recycling and waste reduction.

Some collectors, notably WMI, support legislation that would take the WUTC out of the business of setting rates. This responsibility would go instead to the county, unless the cities had already asserted their jurisdiction. Many smaller trash collectors oppose this change, however. Mindful of the number of small collectors WMI has already purchased, they argue it is important to keep big firms from gaining control over the industry. King County is in the middle of the debate but in general prefers to keep things as they are.

Changes in Risks, Costs, Income, and Power

The county's Solid Waste Division has "profited" from recycling. It is running popular public programs and, until 1991, saw its revenues rise sharply. In that and the following year, however, they declined somewhat because Seattle stopped sending waste to Cedar Hills. There was a further decline in 1993, partly because of the weakening economy and perhaps also because more waste was being recycled instead of being sent to Cedar Hills, which accounts for most of the revenue for the division. As a result, the division may have to cut back the trucking and operations sections slightly. If King County is to achieve its 65 percent goal by 2000, it will have to raise the tipping fees significantly to bring in enough revenues to cover the costs. Higher fees might, however, lead to resistance from cities and demands for greater economy in the division, especially with respect to its practice of setting aside funds for environmental cleanup and the replacement of Cedar Hills. The cities have benefitted from their participation in popular recycling programs and have enjoyed new income from grants and the surcharges they can add to garbage collection bills to support their recycling programs. Neither the cities nor the county has yet faced serious complaints about the rising cost of garbage service.

There is one economic risk from recycling the cities' share: if the markets for recyclables become glutted, the collector has the right to ask that the rates be increased to cover a larger percentage of the cost of curbside collection. If the markets were to collapse and rates skyrocket, cities and counties might face a backlash.

There have been clear losers in industry. Many of the small recycling firms that grew up in the 1970s—the so-called mosquito fleet—have gone out of business. Only a few large firms have been able to afford the capital investment necessary to participate in large curbside recycling programs. Small firms have held on in the commercial waste market by providing recycling services for a few valuable products. However, as the cities and county get closer to the deadlines for the 50 percent, 60 percent, and 65 percent goals, they are likely to push for greater control over the commercial waste stream, that is, to recycle more materials. In the jargon of the system, local governments may want "flow control ordinances" (ordinances that give them the ability to regulate commercial waste). This hot issue is before the state Legislature.

Changes in Administrative Complexity

Recycling and the economic incentives promoting it have made the issues related to waste management facing the cities and the county more complex. Cities have issued separate contracts for the collection of trash, recyclable materials, and sometimes yard waste. Some contracts specify the containers to be used for recyclable materials and apportion the risks between the city and the collector if the market value for recyclables falls below a profitable level. Some suburbs have moved quickly to institute curbside separation and variable fees for trash cans. Several suburbs are establishing curbside programs for multi-family residences; others have not adopted variable rates.

So far, this additional complexity does not seem to have been a political or administrative problem. Instead of writing a simple contract, some cities have issued a request for proposal, a step that shifts the task of sorting through the options and presenting alternatives for consideration to the bidders. Other councils have simply written longer contracts and asked for bids.

Although the county's job is now more complex, it perceives the change as an opportunity rather than a burden. The waste reduction and recycling staffs have new things to do; the operations staff is proud of its new transfer station. Some operations staff would like to set up a county-operated composting service to compete with the new private firms.

EVALUATING THE RESULTS OF THE NEW POLICIES

The 65 Percent Reduction: Early Success and Prospects

So far, King County's campaign to promote recycling and waste reduction has been quite successful: whereas, in 1987, 18.3 percent of the waste stream was being recycled, in 1992 the figure had almost doubled to 35 percent (table 3-4). Improvement may not be so easy in the future, however. Informed citizens and officials generally agree the county and suburbs do not yet have adequate tools to achieve their goal of reducing the waste stream 65 percent by 2000. Most people cite two barriers. One, the markets for several recyclable materials are glutted. There are literally no buyers of green glass, and some recyclers would like to suspend curbside collection. (Some interviewees, however, spoke strongly against doing so, saying it would undermine citizen willingness to recycle.) Most mixed wastepaper is being sold to paper mills in the Far East, but prices are falling. During the summer of 1992, one local recycler paid an Indonesian pulp mill $10 a ton to take almost 10,000 tons of mixed wastepaper off its hands. Other West Coast cities are also exploring the export market for mixed wastepaper. The growing supply could ensure that prices remain depressed.

A second barrier is that the cities have little control over 40 percent of the waste stream because they cannot force businesses to participate in the recycling programs.

A third barrier may emerge—the lack of solid information about the effectiveness of the many steps that have been taken to promote waste reduction and recycling. Little evaluative information is available about most elements of the county's recycling and waste reduction programs. For example, do changes in the rates paid by homeowners encourage them to recycle and reduce waste? There are dramatic differences of opinion about this issue. Most interviewees agreed that participation by homeowners in curbside recycling was clearly influenced by whether the pick-up was a free service. In most cities, customers pay for trash collection and get a recycling pick-up at no extra cost. The interviewees felt that many people would not recycle if there were a separate charge for a recycling pick-up.

There is a great deal of disagreement over the effect of the variable trash can rates. Some interviewees stated emphatically that steeper rates for the pick-up of additional trash cans have encouraged more recycling. Others said the steep rates have just led people to pack more into each can (using the "Seattle stomp"); to dump their trash in a drop box (bin); or to make independent arrangements to get their trash picked up, such as hauling it to transfer stations themselves.

TABLE 3-4
KING COUNTY MIXED MUNICIPAL
SOLID WASTE FIGURES AND PROJECTIONS

Year	Tons Generated	Tons Disposed	Tons Reduced/ Recycled	Percent Reduced/ Recycled
1987	989,500	808,000	181,000	18.3
1988	1,038,500	813,000	225,500	21.7
1989	1,138,500	838,500	305,000	26.4
1990	1,258,500	890,500	368,000	32.1
1991	1,346,500	914,000	432,500	32.1
1992	1,410,000	916,500	493,500	35.0
1993	1,491,500	895,000	596,500	40.0
1994	1,578,000	868,000	710,000	45.0
1995	1,669,500	834,500	834,500	50.0
1996	1,766,000	830,000	936,000	53.0
1997	1,868,000	822,000	1,046,000	56.0
1998	1,976,500	810,000	1,166,000	59.0
1999	2,090,500	794,500	1,296,000	62.0
2000	2,211,500	774,000	1,437,500	65.0
2001	2,332,000	816,000	1,515,500	65.0
2002	2,458,500	860,500	1,598,000	65.0
2003	2,592,000	907,000	1,685,000	65.0
2004	2,733,000	956,500	1,776,500	65.0
2005	2,881,500	1,008,500	1,873,000	65.0
2006	3,038,000	1,063,000	1,974,500	65.0
2007	3,203,000	1,121,000	2,082,000	65.0
2008	3,377,000	1,182,000	2,195,000	65.0
2009	3,560,000	1,246,000	2,314,000	65.0
2010	3,753,500	1,314,000	2,440,000	65.0

Note: The 1991 planning goals forecast has been revised from previous estimates to exclude special wastes (contaminated soils, asbestos, biomedical and industrial waste). The figures for 1987-91 are actual. Those for 1992-2010 are projections.

Source: Draft 1992 Comprehensive Solid Waste Management Plan and EIS, King County Department of Public Works, Solid Waste Division, vol. 1, August 1992, pp. III 2-3.

County council member Barden made one of the strongest statements about the purpose of the variable trash can rates: "They are not an economic incentive. They are a psychological incentive and a political statement. They tell people that it's bad to throw away more trash and that it's good to recycle."[6] Several interviewees pointed out that garbage collection is still relatively cheap and that, although the cost has risen significantly in recent years, it has not climbed anywhere near as fast as the rate for electricity. Most people probably do not pay much attention to the small changes in their bills.

Although the question of whether steeper rate structures encourage recycling might appear to be a central issue in designing a comprehensive solid waste management policy, the county has not studied the matter. The variable trash can rates the county recommended are not based on a formal study; they are set at a level designed to encourage cities to use a steeply ascending rate structure without setting a goal too far out of line with current practice.

So far, the county and suburban cities have not yet felt the need to conduct systematic evaluations of their waste reduction and recycling efforts. They feel no immediate pressure to have better information about which programs work best, as the programs have generally been well-received by the public, the rate of recycling continues to rise, and funds seem to be readily available for waste reduction and recycling programs. However, if movement toward the goal of 65 percent waste reduction or recycling slows, or if the public becomes concerned about the cost of recycling and waste reduction, then better information about the costs and effectiveness of different programs and the nature of the public's response will be needed. Program managers will probably want to know how the variable trash can rates influence the willingness to recycle or reduce waste, the effectiveness of different public education programs, and which locations for and kinds of receptacles for recyclable materials attract the largest volume.

Recent Changes in Information-Gathering and Analysis

Over the past several years, the division has begun to gather much more information than before on various aspects of solid waste management and has upgraded its capacity for analyzing data. When the focus was simply on operating landfills, the information system focused on the revenues raised by the tipping fees. Gatekeepers tallied the overall volume and not the tonnage or nature of the different materials disposed. There was no internal audit.

The new policies have sharply expanded the need for information and analysis. To set tipping fees for Cedar Hills in the context of the facility as a non-renewable asset, the county needs and collects detailed information not only on the current and future costs of operating the facility, including measures to prevent environmental problems, but also on the likely life of the landfill and what a replacement would cost. To comply with the recycling goals of state law, the county and cities also need to know how much trash is being diverted. It is relatively easy to gather information about the volume of materials collected for

---

[6] Interview, February 3, 1993.

recycling. It is harder to know how much yard waste is being composted in backyards or how much waste is simply not being produced because consumers buy different products.

To design a recycling and waste reduction program, the county is gathering detailed information about the composition of the waste stream, obtained by sampling the trash collected. The county also needs information about how many recyclable waste materials can be sold, as the highest priority is to find solid markets for the materials offered for recycling. The county has conducted numerous market studies on recyclable materials—for example, on the amount of newsprint that pulp mills in the Pacific Northwest might purchase and the volume and cost of newsprint imported from other parts of the country.

There is general agreement that more market studies are not needed. What is most needed is information that can be applied to the efforts to intervene directly in the markets. For example, each of the four engineers working at the Clean Washington Center has a small fund for feasibility studies to test new technology for processing recycled materials. The center also works with industry and with programs in other states on engineering studies of the properties of recyclable materials. One study is looking at whether glass contains volatiles or leachates and what its compaction rate might be, another at the performance of shredded rubber in asphalt for use on local highways. Although the director of the center, David Dougherty, is proud of this work, he says that over the long run most testing could be done more efficiently at a national level through a system of federal laboratories or through an impartial industry research program such as the Electrical Power Research Institute, which is funded by electrical utilities.

The center is also investing in information that might make the markets function more efficiently. As Dougherty explains, "The markets for some recycled products are not like commodity markets. There are no consistent definitions of how much contamination is acceptable or of the standard bale size or weight. This means prices fluctuate all over the place."[7] The center is working with the Chicago Board of Trade and major waste management firms to develop standardized specifications; a system for electronic transactions; and an arbitration system for disputes over several products, including newsprint, office paper, and plastics. Some of the brokers who make their living arranging sales of recycled materials think, however, that this effort will lead nowhere; they say they already know the needs of buyers and the standards of the materials available from sellers and that the markets are already quite efficient. What is needed is more data concerning the responses of individuals and households to the program and changes in it.

PROGRAM EVALUATION

The Solid Waste Division is just beginning to look seriously at program

---

[7] Interview, February 4, 1993.

evaluation. Developing a program evaluation system could be at least as challenging as the other analytical tasks the division has undertaken and will demand political acumen as well as analytical skills. Developing a comprehensive system to evaluate the many programs the county and other governments have launched to encourage waste reduction and recycling is a large and complex undertaking.

The key tasks are to identify different programs, specify alternative levels of effort or ways of organizing these programs, and then conduct research to document the cost, the effectiveness, and perhaps the contribution to broad public policy goals (such as the protection of public health) of these alternatives. Realistic analysis of the costs and effectiveness of different programs will of necessity be relatively complex. Some costs will fall on the division, others on various stakeholders. The potential for federal funding, the possibility of recapturing costs through fees, and the incidence of costs on different political jurisdictions or population groups could all be analyzed. For example, the costs of an educational program might be borne by the local school district, university-based developers of curriculum, federal grants, or the business community. The cost of higher fees for picking up trash might fall disproportionately on senior citizens, the poor, renters, or small businesses.

Establishing the effectiveness of different programs is also a complex task, as there are multiple dimensions to effectiveness. Most programs are probably more effective with some elements of the waste stream than others and more effective with some population groups than others. For example, high fees for pick-up might lead to more littering and illegal dumping in some neighborhoods, and fees based on the volume of trash might have little impact on the volume of small amounts of toxic wastes that are routinely dumped into the trash.

Program evaluators might also want to address the broad goals of solid waste management policy, which presumably include reducing the risks to human health and the environment and lengthening the life of valuable assets such as Cedar Hills. Some elements of the waste stream are more of a threat to one of these values than to others. For example, on the one hand, glass is bulky and takes up space in Cedar Hills. On the other hand, in most forms it is relatively inert and poses little threat to human health or ecosystems, provided it is handled and buried properly. In contrast, other wastes that make up a smaller portion of the waste stream may include toxic chemicals that are known threats to human health and the ecology.

In addition to requiring a sensitive analysis of costs, effectiveness, and relevance to broad social goals, a good program evaluation must deal with four additional challenges:

(1) Existing data are inadequate for an effective evaluation, while the cost of obtaining reliable data is high. The waste stream is not tightly controlled in King County, as is probably true in most other jurisdictions. Individuals and businesses can dispose of their trash indepen-

dently, for example, by leaving it in a drop box in an adjacent town. Thus, it is hard to draw connections between rates of recycling and the variable rates for trash cans in different towns. "The single most important factor in explaining how any town is doing is probably the imprecision of the data," says Hansen, the Solid Waste Division's manager.[8]

(2) It is difficult to determine how much a program "reduces" waste. It is relatively clear, at least on paper, how to measure the volume of recycled materials. It is not at all clear how to measure how much trash might have been generated in the absence of a particular program to reduce waste. How can an agency reliably estimate something that did not happen? In the first few years of a program, it is possible to look at whether less trash has been collected relative to the levels of solid waste generated in earlier years. After several years, when multiple programs have been operating, it is much harder to guess how much waste reduction any one program is responsible for.

(3) Individual programs—variable trash can rates, television campaigns, and new kinds of drop boxes and recycling bins—are not independent variables, and it is difficult to sort out the effect of any one program. All the programs operate in the context of widespread public support for recycling, and each contributes to this climate. As one suburban mayor said, "After all of the brouhaha in 1987, people are happy just to go along with recycling."[9] Others noted that recycling was as popular in the Pacific Northwest "as apple pie"; it is part of the region's self-conscious environmental ethic. In this context, it is hard to separate out the effects of specific incentives, messages, or activities. To separate these factors better, intensive interviews with individuals and households are needed.

(4) Public support for the goal of waste reduction and recycling is essential to the effectiveness of many programs. The credibility of the division can have an important influence on public participation in its program. If an evaluation of programs convinced the public that one or two waste reduction and recycling programs were ineffective wastes of money, the result might be a loss of public confidence in recycling in general and less willingness to participate in these programs. Many waste reduction and recycling programs are not instrumental but rather are catalytic in their nature. That is, they do not produce specific products or services for clients, such as traditional trash pick-up, but rather stimulate citizens to change their behavior. They rely for their effectiveness on the willingness of citizens to believe in the information they are given and to feel a moral obligation to participate. Evaluations should not undercut this sense of moral obligation.

---

[8] Interview, February 3, 1993.
[9] Interview, February 4, 1993.

Taken together, these four challenges suggest that where a single local government wants to build its capacity for program evaluation, it should do so incrementally. It might probe for weaknesses in programs that managers suspect to be ineffective and aggressively seek out information about best practices in other jurisdictions that seem to have achieved particularly high overall levels of participation and citizen satisfaction and low levels of waste generation per capita. State and national governments should play a central role as partners in this kind of evaluation strategy. An evaluation strategy needs to be further developed if it is to be clear and useful.

CONCLUSIONS

Four major conclusions emerge from this study. The first is the complexity of designing a system of economic incentives. Some theoretical discussions contrast economic incentives with regulation, as if they were alternatives. In fact, the choices open to managers are much more complex. The management of solid waste in King County uses tools that are usually thought of as economic incentives, notably higher tipping fees and variable trash can rates. These rates, however, function within a system of many other inducements and requirements: King County and suburban cities use a mix of economic incentives, regulation, and other tools such as public education. In addition, governments in Washington also use economic development tools to strengthen the markets for recyclables. The tools here include subsidized research, technical studies, consultation, development and dissemination of information, and assistance in brokering deals between entrepreneurs and financiers to get recycling and waste reduction activities off the ground. These initiatives may be at least as important for recycling as are incentives that are directly reflected in prices, even though theoretical discussions of economic incentives often focus on prices rather than economic development tools.

In addition to selecting an array of tools, managers must decide where to position economic incentives. One key design issue is when to use an incentive in the case of a process that leads to pollution. Public attention has focused mostly on recycling, the second, intermediate stage of the waste cycle. The most important ways to control solid waste may, however, involve the first phase—when the materials are generated—or the third phase—when the wastes are reprocessed into useful products. The practical problems associated with implementing economic incentives in the first phase—which might include taxes on virgin materials and minimum content requirements—are very different from the issues that arose in King County. They may also be much more complex, because there are many more kinds of manufactured goods whose prices might incorporate environmental costs than there are types of trash.

The second conclusion is that economic incentives may not always function as theory suggests. The theory says that the purpose of economic incentives is to levy a financial burden or provide a financial benefit that influences behavior.

Preferences are assumed to be fixed. In a context of fixed preferences, an incentive such as variable trash can rates may influence people's preferences rather than influence their behavior. That is, the most important effects of variable can rates (as well as other tools the King County Solid Waste Division is using) may be education of the public and creation of a psychological and political climate that encourages recycling, rather than influencing calculations about the costs and benefits of recycling. Both approaches, however, are based on the assumption that people behave rationally.

The third conclusion is that the use of economic incentives produces major changes in the management and operations of public agencies. Because of the complexity of the tools used in King County, it is difficult to sort out which implementation issues are attributable to economic incentives and which to related tools. However, three implementation issues deserve special attention:

(1) The use of economic incentives can have a significant impact on agency budgets. Higher tipping fees mean greater revenue and therefore more resources for county recycling programs. Further, increasing the tipping fees and trash can rates may make it possible for cities to hire recycling coordinators with some of the additional funds.

(2) Economic incentives and other new tools will affect an agency's culture. In King County, the use of economic incentives and public education have contributed to a sharp change in the culture of the Solid Waste Division. It is also possible the causation runs the opposite way—as the division has embraced different goals, such as environmental education and changed its clients' behavior, it has hired different kinds of employees who are inclined to use different tools, such as economic incentives and public education.

(3) Economic incentives will affect the total costs and overall efficiency of the solid waste management system. Although the data are incomplete, recycling appears to have raised the costs of solid waste management significantly. Many wastes would not be recycled if separation and collection were not subsidized by trash collection fees. Is this investment wise? It depends. To assess whether an economic incentive is an efficient way of achieving an environmental purpose, the purpose and alternatives for accomplishing it must be defined with care. If the purpose of the solid waste management system is collecting and disposing of trash, then economic incentives to encourage recycling might seem uneconomical also because the costs of solid waste management are increasing. However, if the purpose is to conserve the life of a landfill or prevent toxins in solid waste from getting into the ecosystem via incineration or leaching, then using economic incentives to encourage recycling might seem efficient.

The final conclusion that can be drawn from this case study is the usefulness of program monitoring and assessment. King County monitors the solid waste stream much more closely than it did before 1985. Efforts to assess the usefulness of different devices for accomplishing policy goals are in their infancy, however. This gap has not been a problem so far, because the waste reduction and recycling goals are being achieved on schedule within the range of politically acceptable costs. However, future goals may be more difficult and more expensive to achieve. Information about the costs and impacts of various waste reduction and recycling programs could be of great help in making future policy and management decisions. As King County continues its innovative efforts to address solid waste management issues, the next step may be to design and implement an effective evaluation system.

CASE STUDY:
MICHIGAN'S MANDATORY DEPOSIT LEGISLATION

Deposits on beverage containers are like refundable taxes. The refund value can also be looked upon as a refundable fee charged to purchasers. Those purchasers who throw their containers away pay for the convenience of not returning them, while those who return them, whether they actually consume the beverages or not, are paid for removing bottles from the solid waste stream. In theory, consumers of beverages are paying temporarily to consume beverages in certain types of containers. The deposit is set at a level that does not discourage consumption but that also does not encourage retention or discarding of the containers. To minimize any decline in the purchase of nutritious beverages, the law does not apply to containers of milk, fruit juice, and similar beverages. One advantage of this type of regulation is that society at large is not paying for the costs of disposal, but rather just those consumers who fail to return containers. Those costs of disposal include garbage collection, environmental damage, and wasted energy as a result of using virgin rather than recyclable materials.

The mandatory deposit system is not fully accepted despite this advantage. One argument advanced by critics of the market-oriented deposit system is that deposit laws withdraw beverage containers from the solid waste stream, thereby making recycling programs more costly to operate and more dependent on taxpayer dollars. Beverage containers, it has been estimated, generate as much as 73 percent of total scrap revenues, which offset the costs of recycling programs. Furthermore, if a forced deposit system is instituted in conjunction with a comprehensive curbside program aimed at recycling a variety of materials, people become confused. Returnable beverage bottles often appear in crates along with mayonnaise jars and tomato sauce bottles. Opponents contend dual systems, which is what Michigan has, should be replaced with a single comprehensive recycling system. They often cite Washington's Litter and Recycling Control Law as a model for both litter reduction and voluntary recycling. The bottom line of all recycling programs, they add, should be to recover the maximum amount of waste stream materials at the lowest cost.

It is estimated that up to 95 percent of recyclable beverage containers in Michigan are returned. The return rates for bottles and cans appear to be about equal. Of the nine states that have enacted beverage container deposit laws, not one has repealed the legislation. Nationwide, polls indicate that public support for a national deposit law is strong (box 3-3).

BOX 3-3
ENACTING A NATIONAL BEVERAGE CONTAINER DEPOSIT LAW

In 1991, a telephone survey conducted by the U.S. General Accounting Office (GAO) found that 44 percent of Americans would strongly support and 26 percent would somewhat support a national beverage container deposit law, while 11 percent strongly opposed it and 7 percent were somewhat opposed. In all the states that have enacted such laws, 63 percent of the public strongly approves of them and 19 percent somewhat approves. In April 1993, Peter D. Hart conducted a nationwide poll that showed 76 percent of Americans supported a national deposit law. Support was strongest among younger respondents and those with college degrees.

Despite this approbation, no state has passed deposit legislation since 1983. The nine states that have deposit laws all enacted them between 1972 and 1983. According to Wayne S. Koser, an employee of the Michigan Department of Natural Resources who closely follows bottle legislation, the money spent by the beverage industry to oppose a national bottle bill and state bottle bills has blocked efforts to institute deposit systems. In addition, opponents have falsely but successfully cast the issue as an either/or question between deposit systems and curbside recycling programs. The latter, it should be noted, are paid for by taxpayers, not the industries that sell the products and the consumers that buy them. They also point out that even though many new curbside collection programs are developing, some older programs are being scaled back because of fiscal restraints, with the result that the capture rate for recyclables through curbside collections has declined.

Federal lawmakers have been proposing legislation mandating a national deposit program for more than 20 years. Senator Mark Hatfield (R-Ore.) and then-Representative James Jeffords (R-Vt.) have been the most ardent supporters of a national bill. Until 1977, the legislative proposals sought to ban non-returnable containers. Since 1977, they have called for a national deposit system to encourage recycling. Seventy senators and representatives are sponsors of the Beverage Container Reuse and Recycling Act of 1993 (H.R. 1818 and S.818), which calls for each state to recycle 70 percent of its beverage containers. That rate is, however, well below those achieved in states with deposit legislation. If a state cannot reach the 70 percent rate two years after enactment, a mandatory deposit of 10 cents would be required on all beer and soft drink cans and bottles. Committee hearings have been held on the bills on a number of occasions, but they have never been reported out.

Proponents of a national law include the League of Women Voters of the U.S., the U.S. Public Interest Research Group, environmental organizations and other non-profits. They have banded together to form the National Container Recycling Coalition. The coalition argues that a national bottle law would reduce litter, protect health and the quality of life, encourage recycling, and end the financial burden of costly waste disposal systems.

Opponents reply that national legislation offers only a partial solution to

continued on next page

BOX 3-3 *continued from preceding page*

the problem of waste disposal because beverage containers constitute only 3-5 percent of the total waste stream. They also argue a national law would be costly and unfair to the beverage industry. Container, beverage, retail trade associations and other organizations opposing the bill have formed the Coalition Against Forced Deposits. The contrasting views of advocates and opponents of national deposit legislation are shown in the table below.

To understand why national bottle bills have not been reported out of a number of congressional committees and have not been debated on the floor of either chamber, the dynamics of the legislative process, particularly the influence of lobbyists, needs to be understood. The beverage industry, supermarket chains, and unions for steel, glass, and aluminum workers have been contributing funds to oppose deposit legislation and mobilizing grassroots lobbying against mandatory deposit bills. In contrast, the pro-bottle bill lobby is a loose coalition of civic groups with limited financial resources.

Topic	Advocates	Opponents
Solid waste reduction	Solid waste and litter would be reduced significantly, especially on highways.	Solid waste would not be significantly reduced; beverage containers make up, at most, 5% of the waste stream.
Energy conservation	Energy savings would be realized because of a consistent supply of recycled materials.	Energy savings would be higher if comprehensive recycling programs were in place.
Recycling of other materials	Recycling of other materials would occur, because deposit laws change behavior.	A costly network would be created for the recycling of other materials
Litter reduction	A mandatory bill would reduce beverage container litter, cut the total volume of litter by 30-60%, and cut cleanup costs in half.	A national deposit bill would not regulate the biggest contributor to litter–plastic foam containers.
Danger to humans and animals	Broken bottles would continue to damage bicycle and car tires, farmer's equipment and injure livestock	
Tourism	Cleaner parks and highways could attract tourists.	
Employment	The bill would create thousands of jobs in the beverage, retail, transportation, and recycling industries.	The bill would eliminate skilled jobs in the glass and can manufacturing industries.
Prices and sales	A national law would result in increased use of refillable containers, which cost 35% less than containers used once, and so will attract customers back to stores.	A mandatory law would lead to increased prices and decreased sales as a result of increased handling costs.
Sanitation	There have been no reported sanitation problems as a result of state mandatory deposit laws.	Sanitation problems would result at retailers' or dealers' locations.

Source: National Academy of Public Administration, 1993.

Before Michigan enacted its 1976 law establishing mandatory deposits on beverage containers, curbside recycling initiatives, and other programs to reduce solid waste, beverage containers were a highly visible and costly source of litter in Michigan. Bottles and cans were strewn along the roadways, in parks and recreational areas, along city streets, and in the countryside. Michigan began implementing its bottle law in 1978. The goal was to reduce this source of litter, as well as to conserve energy and natural resources.

This section reviews Michigan's experience with its bottle law.[10] The next part looks at the origin of the Michigan law and how it works. The second part discusses the results of the law and the costs and administrative burdens associated with its implementation. A number of performance indicators for evaluating the program are discussed in the third part. The section ends with a number of conclusions concerning the implementation of Michigan's deposit program.

This case study involved interviews with state officials, recyclers, bottlers, grocers, and other key informants to collect data on roles and responsibilities, manpower requirements, activities, perceived costs and benefits, and issues arising in implementation. It also involved a review of the literature on the Michigan bottle bill. It was impossible to conduct systematic research into the effectiveness and efficiency of Michigan's program, however. Formal evaluations have not been conducted; performance monitoring systems are non-existent; and the private sector is not required to provide data on the quantity of containers recycled, the costs of handling the containers, and the amount of unclaimed deposits. How well the law has responded to considerations such as the spread of costs among various groups, including bottlers, distributors, and retailers, and how well it facilitates adaptation to changing circumstances have rarely been addressed.

THE BOTTLE BILL

In 1976, Michigan voters adopted the Mandatory Beverage Container Deposit Law (MCLA 455-571), the "bottle bill." The law, which became effective on December 3, 1978, stipulates that:

- All containers for carbonated beverages (including alcoholic beverages but excluding beverages containing wine) require a 10-cent deposit, except that certain certified beverage containers require a 5-cent deposit. (A beverage container is certified if it is reusable by more than one

---

[10] Michigan was chosen for a case study for several reasons. As noted, there is strong public support for deposit programs. At the same time, there is a great deal of interest in market-oriented programs within the environmental arena. Thus, it seemed appropriate to do a case study of a deposit program. Michigan was a logical choice because its program has been in operation since 1978, so that its experience is relatively long. Michigan was also appealing because its program relies almost exclusively on the private sector. The role of government is minimal—on a continuum of public sector involvement, Michigan's deposit law falls at the end, where governmental activity is practically non-existent.

manufacturer in the ordinary course of business and if more than one manufacturer will accept the beverage container for reuse and pay the refund value of the container.)
- Retailers and distributors accept returned beverage containers and refund the deposits.
- Non-returnable containers and detachable pull tabs on cans are banned.
- The name of the state and the refund value must be printed on all except specially exempted containers.

The law can be amended or repealed only by the voters or by three-fourths of the members of both the state Senate and House of Representatives.

Passage of the referendum was attributed to the ability of a coalition called Help Abolish Throwaways to raise public awareness by skillfully using the mass media. Because of the "fairness doctrine," both proponents and opponents of the legislation had access to radio and television. That doctrine, first applied in 1949, imposed a two-part obligation on broadcasters. First, they were required to provide adequate coverage of controversial issues of public importance, and, second, they were required to provide reasonable access to opposing viewpoints. The significance of this requirement for equitable coverage should not be underestimated. In addition, the *Detroit Free Press*, which is read throughout Michigan, enthusiastically supported the initiative. It provided ample coverage of the issues and ran several editorials endorsing the bill. Finally, the $72,000 contribution from the Michigan United Conservation Clubs paid for an important last-minute media blitz. (See box 3-4 on the campaign to get the bill passed.)

In 1977, the law was amended to include deposits on refillable soft drink containers manufactured before December 3, 1978. It was amended again in 1986 to place mandatory deposits on wine coolers. In 1989, it was further amended to create an unclaimed bottle fund; the fund was to be used for environmental improvements and to be administered by the state. That amendment is still under appeal.

Given the Michigan law's overwhelming margin of victory in 1976 and continued popularity, it is not likely to be repealed. Opponents and proponents both believe the majority of citizens think it is sacred. Opponents admit the program has reduced litter, the number of personal injuries from broken bottles, and damage to farm equipment.

How the Bill Works

The Michigan bottle law requires, as noted, that consumers pay a 10-cent deposit on all containers of carbonated beverages (soft drinks, soda water, and mineral water) and on malt drinks sold in airtight containers of one gallon or less, and a 5-cent deposit on certified beverage containers. The containers must have the refund value and name of the state marked on the container. Retailers

BOX 3-4
HOW MICHIGAN'S 1976 BOTTLE BILL GOT PASSED

The first attempt to regulate beverage containers in Michigan came in the early 1960s, when the Michigan State Liquor Control Commission recommended a ban on one-way non-returnable containers to reduce litter and counteract other effects of what the commission characterized as a "wasteful" society. Michigan's Stroh Brewery supported the recommendation because it was attempting to end the advantages held by other brewers that were marketing one-way containers in Michigan. However, then-Governor John Swainson requested a postponement of the effective date of the ban to allow for further review. The ban never took effect, and by the 1970s, one-way glass containers and cans accounted for a major share of the beer and soft drink markets. All deposit legislation introduced in the Michigan Legislature from 1965 to 1976 met the same fate, never even coming close to passing.

When the 1976 bottle bill was proposed, it generated lively debate. In light of the prior unsuccessful attempts to get bottle bills passed, this time numerous environmental and conservation organizations, including the Michigan United Conservation Clubs (an organization that represented more than 400 local conservation clubs and sportsmen's groups and boasted more than 100,000 members) and the League of Women Voters, got together and formed the Help Abolish Throwaways coalition. Emphasizing the potential benefits, such as reduced litter from containers and lower disposal costs, preservation of natural resources and energy savings, the coalition had no problem recruiting volunteers to help with a petition drive to place the measure on the statewide ballot. It distributed petitions through the magazine of the Michigan United Conservation Clubs and newsletters on agriculture and conservation. Governor William G. Milliken endorsed the bill, signed the first petition and paved the way for its endorsement by several state agencies.

Opponents of the bill included the beverage container manufacturing industry and producers of beverages, distributors, retailers, and labor unions. They argued it would be discriminatory to ban one-way beverage containers because of the adverse effect on the beverage industry and that deposit legislation would result in lower employment, lower state taxes and revenues, and higher beverage prices. They contended that a more comprehensive resource recovery program would be less restrictive and more efficient than a reduction program aimed at a single source. Such a resource recovery program would preserve consumer choice while reducing many kinds of solid waste, energy consumption and roadside litter, as well as unemployment. The opposition formed the Committee Against Forced Deposits and financed a major media campaign to defeat the issue.

The Committee Against Forced Deposits raised $1.3 million to campaign against the bill. The Help Abolish Throwaways coalition raised only $45,000 in contributions. The Michigan United Conservation Clubs made a $72,000 contribution from its general fund to supplement the $45,000, however, for a total of $117,000. The conservation coalition, well-organized and strongly led, pursued a successful strategy to obtain significant grassroots support throughout the state. During the six-week petition drive the coalition got 400,000

continued on the next page

144 *The Environment Goes to Market*

> **BOX 3-4** *continued from preceding page*
>
> signatures, nearly double the number required to place the issue on the ballot. Of those who voted on the initiative, 64 percent approved it, and only 3 of 83 counties disapproved of it.
>
> Thomas L. Washington, president of the Michigan United Conservation Clubs, pointed out one of the reasons for the widespread support. One-third of Michigan's economy was based on industrial production, one-third on agriculture, and one-third on tourism. Litter adversely affected the tourism and agricultural bases. Because disposable containers affected two of the state's three economic foundations, supporters were fairly optimistic the bill would get passed.

that violate the law are subject to a fine of not less than $100 and not more than $1,000 plus court costs for each day they are in violation.

On average, a container may change hands three or four times from producer to consumer. In the beer industry, deposits on refillable bottles generally originate with the brewers, who receive a 10-cent deposit on each container when the distributors pick up the containers. The brewers, in turn, refund to their distributors the deposits for empty containers the distributors return to them. Anheuser-Busch deviates from this practice in that it begins the chain of deposits on cans as well as bottles. In contrast, the Miller Brewing Co. initiates deposits only on refillables; its distributors start the deposits on other beer containers, including cans. In the soft drink industry, the bottlers usually act as both producers and distributors of beverages. The deposits start with the bottlers, who collect the deposits from the retailers. The bottlers charge their retailers the amount of the deposit on all beverage containers when they deliver the containers to their retailers and refund the retailers the deposits on all empty containers they get back from them. Consumers, in turn, pay the retailers a dime when they buy a beer or soft drink. When consumers or others return the bottles, they reclaim the deposits. When consumers do not return the bottles, the bottlers retain the unclaimed dimes. (Figure 3-2 shows how this system works.)

The advantage of initiating the cycle is obvious: retaining the unredeemed deposits. In this respect, Anheuser-Busch fares better than Miller. However, Miller distributors offset their handling costs by collecting the unclaimed deposits on all bottles and cans except refillables. Miller distributors also obtain revenue from the scrap derived from returned bottles and cans they recycle. Initially, Anheuser distributors received no revenue from Anheuser to offset handling expenses. In 1979, however, recognizing that the additional costs incurred by its distributors put them at an unfair advantage among distributors, Anheuser began to pay the distributors 1 cent per container returned. The law does not require that the manufacturers pay handling fees to distributors or that distributors pay handling fees to retailers.

In short, the consumer pays a dime to a distributor or bottler every time he or she breaks, discards, or loses a returnable container. The retailers, as

FIGURE 3-2
HOW THE MICHIGAN DEPOSIT-REFUND SYSTEM WORKS

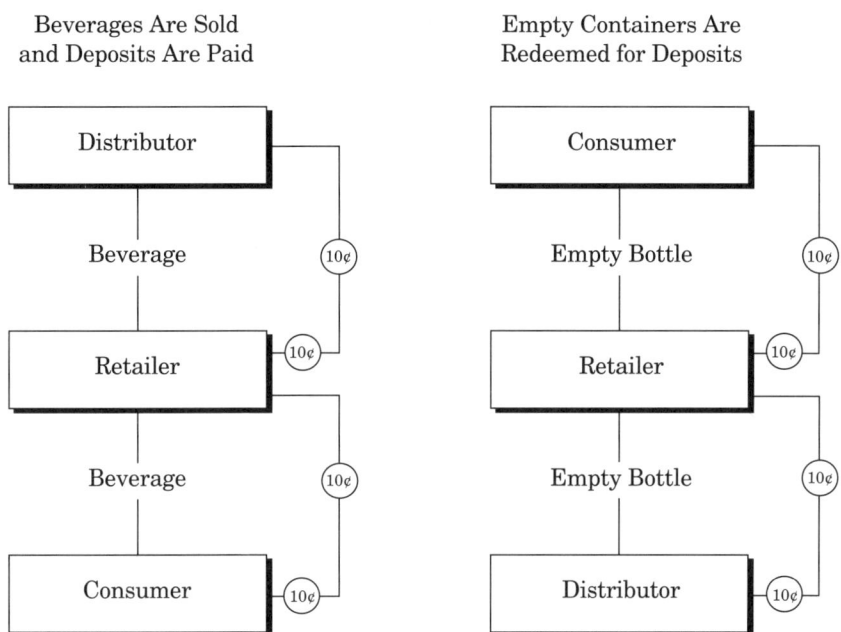

Source: "Tradeoffs in Beverage Container Deposit Legislation," GAO/RCED-91-25, p. 17.

middlemen, do not benefit from unclaimed deposits because they transfer the dime deposits from the consumers to the distributors or bottlers. In the minds of the public, the deposit process is straightforward. In the minds of the bottlers, distributors, and retailers, it is not, even though the procedures are relatively routine.

Unclaimed Deposits

Unclaimed deposits on unreturned containers affect bottlers and distributors as well as consumers. Consumers lose money equal to the sum of the unclaimed deposits. The profits derived by distributors and bottlers are equal to the losses of the consumers. Unfortunately, under Michigan law, distributors, brewers, and bottlers are not required to keep records and report the amount of unclaimed deposits, including profits. The exact amount of unclaimed deposits is therefore unavailable. According to a number of estimates, including those of the GAO and U.S. Department of Commerce, the annual loss to consumers is between $20 million and $52 million. This range includes unclaimed deposits on all beverage containers covered by the deposit law.

Not surprisingly, bottlers and distributors contend the unclaimed deposits amount to $20 million a year. In 1988, the Michigan Soft Drink Association asked Temple, Barker and Sloane, Inc., a Massachusetts consulting firm, to determine all costs resulting from the deposit law (costs include investments in facilities, recycling equipment, expanded warehousing, additional labor, and new trucks to accommodate the law).[11] The firm determined the 1988 rate of return for soft drink containers was 95.3 percent and the unclaimed deposits were $9.1 million. The net costs to the soft drink bottlers of collecting, storing, and recycling the bottles came to $14.2 million. William Lobenherz, president of the Michigan Soft Drink Association, argued that industry pays taxes on the unclaimed deposits just as it does any other revenue and that the profits from unclaimed deposits do not begin to compensate for the capital expenditures such as pallets and washing machines associated with implementing the law.

While such arguments have some validity, taxation is less important than the issue of whether or not industry is entitled to the profits from unclaimed deposits and the returns on investments from these profits. According to Public Sector Consultants, Inc., a Michigan consulting firm, if an estimated $48.6 million in deposits is unclaimed annually and 80 percent is invested (the rest is used to defray costs) at a market interest rate of 6.1 percent, companies could earn $2.4 million in interest income a year.[12] (This figure assumes the return of 434 million bottles and cans annually and an average deposit of 11.2 cents per container.)

Since 1979, Michigan United Conservation Clubs has supported legislation that would require industries to report the profits attributable to unclaimed deposits and turn 75 percent over to a revolving state conservation and recreation fund administered by the Michigan state treasury. The remaining 25 percent would be allocated among bottlers, distributors, and retailers to offset the costs of complying with the bottle law. The allocations to each of these dealers would depend on the number of empty returnable containers they handled. Advocates of this proposed law were encouraged by the state of Maine, which several years after passing deposit legislation amended it to channel deposits into a state fund. The Maine amendment survived a court challenge. In 1989, Massachusetts passed a similar amendment, specifying that the unclaimed deposits, amounting to approximately $21 million a year, would be put into a fund dedicated to recycling programs. It was rejected by a lower court but then upheld by the Massachusetts Supreme Judicial Court in March 1993.

THE IMPACT OF THE BOTTLE BILL

Michigan's container deposit program has proven highly successful: the return rate is estimated at 93-95 percent, compared with 79-83 percent in other

---

[11] Temple, Barker and Sloane, Inc., "Michigan Soft Drink Bottlers' Costs Resulting from the Deposit Law," prepared for the Michigan Soft Drink Association, Lexington, Mass., April 1989.

[12] Mark Jorritsma, "Unreturned Beverage Containers and Unclaimed Deposits," *Public Policy Advisor* (Public Consultants, Inc.), August 25, 1986.

states. Those interviewed credit the 10-cent deposit for Michigan's exceptional rate: all other states with deposit programs have a 5-cent deposit. However, a 1983 study[13] concluded the size of the deposit has little effect on the rate of return, at least for deposits of 2 cents or more.

The Outcomes

The outcomes of the Michigan bottle law are as follows:

- **Less litter.** According to the Michigan Department of Transportation, before passage of the law, beer and soft drink containers constituted 17 percent of all roadside litter. Eight years after the law took effect, the rate was 4 percent. Several studies cite a 78 percent reduction in cans and 51 percent reduction in bottles.

- **Less solid waste.** According to the Michigan Department of Natural Resources, 6-8 percent of the state's solid waste (more than 600,000 tons annually) is diverted as a result of the bottle law. This reduction in solid waste has produced additional savings in reduced labor and equipment, in the extension of the useful life of landfills, and in the lessening of corrosion and abrasion of energy recovery incinerators.

- **More jobs.** In 1988, the GAO estimated that the deposit program created 4,888 jobs; 720 soft drink bottlers; 68 brewers; 600 beer distributors; and 3,500 retail employees. In contrast, 240 jobs were lost, primarily in the retail can and glass manufacturing and litter and waste disposal sectors. The net change amounted to an estimated increase of 4,648 jobs. The new jobs were created all along the beverage distribution chain; at the retail level they were greatest in the metropolitan areas. On the other hand, the jobs created were low-paying and unskilled.

- **New technologies.** The program reportedly stimulated the development of innovations in can crushers and container sorters and the creation of products from recycled materials. For example, to relieve storage problems, Detroit Coke developed a plastic bag for cans with a 10-case capacity.

- **Energy savings.** The manufacture of cans and bottles from recycled materials produces energy savings. According to one study,[14] Michigan saved an estimated 8.1 trillion British thermal units (Btus) in 1984 as a result of the bottle bill. This 8.1 percent reduction in Btus amounted to about 60 percent of all the energy used by the bottling and related industries in 1975. Another study examined the total Btus used in production, including mining the raw materials, transporting them,

---

[13] Richard Porter, "Michigan's Experience with Mandatory Deposits on Beverage Containers," *Land Economics* 59 (May 1983):177.

[14] Op. cit., Jorritsma.

transforming them into containers, washing and refilling containers, and transporting them to stores.[15] According to this study, before the bill took effect, 55 million Btus were required to produce 1,000 gallons of beer. After the law was implemented, only 34 million Btus were required to produce the same amount, a 37 percent reduction in energy consumption.

Overall, refillable bottles are the most energy-efficient containers. Plastic bottles, their nearest rival, use twice as much energy. Most of the energy spent in the use of any container is petroleum and natural gas. The refillable bottle is most efficient with respect to both these sources. Although the fuel consumed by the trucks that collect the containers and by the sorting and crushing equipment offsets some of the production savings, there is still a net energy saving.

- **Changes in environmental policies.** After passage of the bottle bill, the Michigan Natural Resources Commission (NRC) adopted a State Resource Recovery Strategy that set a recycling goal of 25 percent of all solid waste by 1995. The NRC began a "Buy Recycled" campaign. Legislation was enacted and executive orders were issued to encourage recycling, and the number of recycling centers was increased. While state employees claim that curbside programs and the deposit system complement one another, no data have been produced for Michigan supporting their claim. However, a report released by the Congressional Research Service (CRS) in January 1993 concluded that curbside recycling and deposit systems are compatible and that the combination of the two would result in increased diversion rates and higher quality collection. (See box 3-5 on the issues that generally arise in conjunction with the adoption and implementation of bottle bills.)

- **Health and safety.** Lacerations caused by broken glass and tab tops have been almost entirely eliminated, and farmers report that damage to machinery has declined. Grocers, however, report that soiled and contaminated containers have caused sanitation problems in storage areas and that they must bear the costs of maintaining the storage areas. This point raises an equity question with respect to the allocation of the costs associated with operation of the deposit system.

- **Packaging.** There has been a shift from cans and one-way glass bottles to refillable glass bottles and plastic bottles, mainly two-liter. The shift has been less pronounced in the beer industry than in the soft drink industry. The main factors influencing packaging decisions beside the deposit law are the prices of the containers, consumer preferences, and the ability of consumers to pay.

---

[15] Richard Sjolander and Peter Kakela, "Effects of Michigan's Mandatory Beverage Containers Deposit Law," Michigan State University, East Lansing, Mich., 1984, p. 65.

> **BOX 3-5**
> **ISSUES THAT GENERALLY ARISE WITH BOTTLING BILLS**
>
> - Changes in employment.
> - Rights to the unclaimed deposits.
> - Allocation of the revenues from unclaimed deposits.
> - Costs to bottlers, distributors and retailers.
> - Energy savings.
> - Compatibility with other recycling programs.
> - Impact on sales of beverages.
> - Convenience.
> - Impact on other regulatory programs such as comprehensive recycling.
> - Packaging changes.

Source: National Academy of Public Administration, 1993.

- **Product prices.** In the year after passage of the bottle bill, beer and soft drink prices rose 10-15 percent, or $102 million-$164 million. However, after the capital investments associated with the law were completed, prices fell by an equivalent amount. Subsequently they rose again, going from $102 million overall to $164 million annually from 1979 to 1981, or between 2 cents and 5 cents per bottle filled. From 1982 to 1988, the increase was from $9 million to $94 million annually, or between 0.2 cents and 2 cents per bottle filled. It is difficult, however, to calculate precisely what portion of the increases and decreases are attributable to the bottle bill, given the many variables that affect beverage prices, ranging from the rate of inflation to advertising expenses to the popularity of other beverages such as coffee.

- **Product sales.** Figures that permit a comparison of purchases of soft drinks and alcoholic beverages in Michigan before and after passage of the bottle bill are not readily available. It is estimated that shortly after the bill was passed, soft drink sales fell by 5-10 percent and that during 1979, the law's first year, beer sales dropped by 3.6 percent. However, many believe other factors that arose at about the same time the law took effect—a decline in tourism, a rise in unemployment, higher beer prices, and a change in the drinking age from 18 to 21–contributed to the drop-off in sales. All that can be said with assurance is that in 1990 Michigan ranked in the middle nationally with respect to the purchase of alcoholic beverages—23rd in spirits, 27th in wine, and 28th in beer. In addition, per capita beer consumption hardly changed between 1981 and 1991– 23.32 gallons versus 23.29 gallons per person annually. Again, factors other than deposit laws influence consumption, including income, health consciousness, and changing demographics.

150   *The Environment Goes to Market*

- **Non-profit recycling centers.** Michigan's non-profit recycling organizations lost potential income as a result of the decline in aluminum and glass recycling. However, a number of community recycling organizations continue to earn a small fraction of income from containers discarded by individual households. Several interviewees reported that retired and homeless people often collect bottles to supplement their income. Thus, they have benefitted from the bottle bill.

- **Purchasing patterns.** It was reported that citizens from cities and towns near the state's borders initially purchased their beverages in neighboring states after the law took effect. Interviewees said the reason was increased prices and the higher minimum drinking age in Michigan. During the first few years after the bill took effect, it was also claimed that drinks such as fruit juices, which are not covered by the law, achieved a larger market share. However, there is no evidence of that change in purchasing patterns.

Costs

Distributors' and Bottlers' Costs

The Michigan Soft Drink Association estimates that in the first two years of the deposit program soft drink bottlers made $93.9 million in capital investments. Current capital costs in 1988 were $5.5 million and consisted of additional vehicles, forklifts, baling equipment, bottler washers, and bottle inspection equipment. The bottlers estimated that operating and maintenance costs, including utility costs, amounted to $9.4 million, and labor costs accounted for $24.2 million. Of the labor costs, additional drivers accounted for the largest share (36 percent), followed by added help to operate the bottle washers and refillable bottle filling lines (35 percent) and to unload, sort, and move empties before sale to scrap buyers (28 percent). Thus, in 1988, the total cost of implementing the legislation to the bottlers alone was $39.1 million. The bottlers offset these costs in part with revenue from the sale of scrap, principally aluminum, and from unclaimed deposits. The net costs came to $14.2 million (table 3-5).

In 1989, the Michigan Department of Natural Resources came up with quite different figures. It estimated the costs to bottlers and distributors in 1988 were $70 million but that the revenues from the scrap and unclaimed deposits totaled $113 million-$118 million. The net gain to bottlers and distributors was $43 million-$48 million.

Detailed information on the costs the legislation has forced bottlers to bear is unavailable. It should be noted that the costs decline once the initial capital investments in facilities, equipment, and transportation have been made.

Retailers' Costs

Retailers incur costs under deposit laws because they must collect, sort, store, and redeem the containers. For example, one grocery store chain reported

TABLE 3-5
COSTS BORNE BY SOFT DRINK BOTTLERS, 1988

Bottlers' total costs	$39.1 million
Scrap sales revenue	-15.9 million
Unclaimed deposits	-9.1 million
Total	$14.2 million

Source: Temple, Barker and Sloane, Inc., "Michigan Soft Drink Bottlers' Costs Resulting from the Deposit Law," prepared for the Michigan Soft Drink Association, Lexington, Mass., April 20, 1989, p. 9.

spending close to $1 million on storage facilities, sanitation services to control pests in storage areas, accounting forms, and container carts. Another chain said it spent $2 million on similar items. Although data on these costs cannot be verified, all interviewers acknowledged that retailers bear a heavier burden than any other segment of the beverage sales industry. They cannot take advantage of the potential means to reduce or offset the costs that other levels of the beverage industry have: as noted, they do not retain unclaimed deposits because they do not originate the deposits, nor do they hold deposits; they do not end up with any materials they can recycle; and they are not affected by changes in packaging.

In Michigan, the costs to retailers are not offset by handling fees paid by distributors to retailers as in several other states with mandatory deposit legislation. A representative of the Michigan Grocers' Association noted that legislation has been introduced to provide a handling fee to retailers, but the amendment has not passed. Retailers, however, do not believe giving them a percentage of the unclaimed deposits is an effective way to compensate them for the handling costs; they want a formula that compensates them on a per container basis, as many redeem far more empty containers than they sell.

When asked if the bottle law has affected store traffic and relations with customers, a representative of the Grocers' Association observed that store traffic appears to have remained the same. However, at certain large stores customers must wait in line to return bottles during peak periods. This wait creates frustration and hurts consumer relations. Grocers do not believe the bottle bill has led to increased sales as some originally anticipated—they contend that "redeemers" are not converted into customers.

Grocers report an increase in friction between distributors and grocers. During pick-ups, tempers may flare. They note that keeping storage areas clean is very expensive and that there have been sanitation problems in storage rooms. One reason is that consumers do not clean containers the way they used to clean milk bottles, and the cost of baling the cardboard containers is high.

Michigan has not established redemption centers, which in certain states alleviate the sanitation concerns and interruptions to business. The law allows for the establishment of centers but does not provide financing for their construction and operation. In 1985, Michigan did limit the number of containers retailers must accept from any one customer during one day to $25 worth as a way to reduce the logistics and storage problems.

Overall Costs

One study on the economic effects of the beverage container law concluded that it is a very expensive way to reduce litter.[16] He quotes Peter Stroh of Stroh Breweries as saying the bill is a $60 million solution to a $3 million problem—in 1977, Stroh claimed the costs of handling a returnable system amounted to $60 million a year, whereas the state spent $3 million on litter pick-up. In 1978, Murray Young of the Associated Food Dealers of Michigan claimed the cost of implementing the bill was $250 million whereas the cost of litter pick-up was $10 million.

No one knows if these figures are accurate. However, the initial costs of the law were probably high, because they included a one-time investment in capital equipment, and the costs associated with implementing the bill are not trivial.

Administrative Burden

The deposit program imposes almost no administrative burden on the state. The marketplace is left to oversee the law; no overall administrative responsibility such as monitoring or enforcement falls on public agencies. The bill is, in effect, self-enforcing. Unlike many command-and-control-style regulations, deposit laws require no research, standard-setting, or monitoring, and enforcement is limited to acting on violations reported by consumers or other parties involved in beverage transactions. How little the state agencies have to do can be seen in the following discussion.

The deposit law states that the Michigan Liquor Control Commission must certify all containers. Any container certified as refillable by more than one manufacturer requires a 5-cent instead of a 10-cent deposit. As of 1993, there were no certified soft drink containers with a 5-cent deposit, and only a limited number of beer bottles of various sizes had been certified for the 5-cent deposit. A single employee who does not need special training handles the certification.

For more than a decade, the Michigan Department of Natural Resources has furnished information on the bill to researchers and officials from other states; the individual responsible for outreach estimates he spends approximately 40 hours a year on external relations. Expenses for the duplication of written information on the law are a couple of hundred dollars a year. When asked whether the limited role of the department resulted in an inability to integrate the deposit program successfully with other environmental programs, such as residential recycling, a department representative maintained there have been no adverse consequences.

---

[16] Richard Sjolander, "The Economic Effects of the Michigan Beverage Container Law on a Package System," Master's thesis, School of Packaging, Michigan State University, East Lansing, Mich., May 31, 1975, p. 9.

The Michigan Department of Transportation has conducted three studies of roadside litter since the introduction of the bottle bill, the last one in 1986. Although the studies were originally designed to assess the impact of the bill on costs and other factors associated with the reduction of litter, they document and measure changes in the composition of roadside litter. The studies concluded that bottles do not comprise a significant portion of roadside litter. The department has the skills necessary to perform such studies, but a lack of resources makes further ones unlikely.

If a retailer refuses to accept a returnable container, the customer may file a complaint with the attorney general's office. This office reportedly sends about half a dozen form letters a year to retailers that refuse to accept returnable bottles. The letter warns of fines of $100-$1,000 for each day that retailers are in violation plus the costs of prosecution. The law clearly spells out the penalties, so that there is virtually no room for administrative discretion. The administrative costs associated with enforcement are negligible.

Under the Michigan law, distributors, brewers, bottlers, and retailers are not required to keep records of how many beverage containers are returned, let alone report the profits they make from unclaimed deposits. Several interviewees noted that in essence the private sector administers the bottle bill (the bill was designed so that taxpayers would not pay the administrative burden).

It should be kept in mind that the objective of the Michigan bottle bill was to reduce litter and solid waste with minimum losses to businesses, minimum inconvenience to consumers, and minimum cost and administrative burden to the government. When the bill was passed, voters did not seriously consider other options for reducing litter, such as universal product charges; clean community systems; prohibitions on particular kinds of container materials; or public education programs on litter, litter law enforcement, or buy-back recycling programs. They simply believed a container deposit program would provide the most effective means of reducing litter. Moreover, the bill was considered in isolation. That is, its relationship to other environmental programs, such as materials recovery, was not taken into account.

Despite the sympathetic attitudes of public employees toward the law, interviewees, including state officials, industry representatives, and policy analysts, expressed increasing concern about the lack of adequate and up-to-date information on its implementation. There are no mechanisms for collecting the information that would enable the program to be monitored and evaluated, the impact of the law to be assessed, and the program to be coordinated with other environmental policies. Employees of the Department of Natural Resources pointed out that it is impossible to determine the impact of a number of variables that affect the deposit program because data are not available.

Representatives of several organizations that supported the bill believe the provisions relating to unclaimed deposits need to be revised. Asked if there is flexibility to change the bill, public employees remarked that such a proposal has already touched off a major political battle. The bill is as rigid as, if not more rigid than, a command-and-control instrument. They speculated that because the original bill was not originally conceived as a revenue bill, it may fail to become one. Although the public is increasingly aware of the need to protect the environment, the beverage industry is likely to block changes, for example, that would put deposits in a special fund earmarked for environmental protection or to authorize handling fees for distributors and retailers. The proposed changes will be difficult to achieve because they will not advance the interests of manufacturers and distributors who continue to believe that forced deposits should be scrapped. The lesson for advocates of deposit laws is that they should think through the details when drawing up legislation—once on the books, the laws may prove as hard to change as Michigan's has been.

EVALUATION

Although there has been no systematic evaluation of Michigan's mandatory deposit program, the system lends itself to the generation of information, statistics, and performance indicators. There are three types of performance indicators:

- **Input.** Input indicators generally answer questions about the size and scope of the program and whether it has changed over time. An example of an input indicator is the materials used in containers in Michigan.

- **Process.** Process indicators describe the services, activities, and employees involved in operating the system. Examples of processes are the collection, storage, cleaning, crushing, and transportation of containers.

- **Outcome.** This cluster of indicators describes the changes that have resulted from the deposit system. Examples are measures of litter and solid waste reduction, effects on health, beverage sales and consumption patterns, and revenues from unclaimed deposits.

Performance indicators can be used in three management areas: reports to the Legislature, policy decisions, and operations. Specifically, indicators form a basis for comparisons before and after initiation of the program, showing how the program has changed over time and how it rates against norms, standards, and objectives.

Performance indicators are just one piece of an evaluation system or feedback loop. An entire evaluation system would determine (1) whether and how well the goals of the system are being met and whether the goals are still desirable or should be changed (goal-setting), (2) performance, (3) monitoring, and (4) causal effects of the program.

Some data sources for indicators are currently available. Others can be obtained by designing standardized reporting forms to obtain information, by using systematic coding methods, and by interviewing stakeholders. Unfortunately, data for the component of program evaluation intended to provide causal information are impossible to gather at this time. Data on cause and effect should have been considered in advance of the deposit system's implementation rather than retrospectively. Doing so would have made it possible to make plans for and to select appropriate methods to infer causality and to control for the influence of a number of variables on particular outcomes. Usually information relating to causality is obtained through experimental or quasi-experimental designs.

To develop a proper set of performance indicators useful to policy-makers and stakeholders such as bottlers, distributors, retailers, and recyclers, it is advisable that state employees and representatives of stakeholder groups meet to (1) suggest key indicators, (2) rank the indicators in order of importance, and (3) reduce the number of indicators to a manageable number. The final indicators should be checked against criteria for inclusion—for example, is the indicator an important source of information? Can it be collected at a reasonable cost? Is it reasonably accurate? The indicators may be classified in a number of ways. For purposes of this discussion they are classified as input, process, and outcome. In any event, those administering the evaluation and those who will use the results should review the final set of indicators. The proof of the pudding will come when the indicators are used to measure and compare, when reports are made public, and when changes are made in the system.

Successful evaluation of the program requires that attention be paid to the creation of an institutional structure, perhaps in the Michigan Department of Natural Resources. Regardless of where the evaluation function is housed, institutional interests and credibility are involved. A staff member with a minimum of a master's degree and experience in instrument design and data systems is needed to design the instruments and indicator reports, collect and process the data, analyze the results, and publish the reports. Of particular importance is the need for clear instructions to avoid methodological problems, such as double counting, and to establish responsibilities and deadlines, so as to facilitate the data collection and analysis.

The following provide a starting point for the development of a system of performance indicators for Michigan's deposit system (the actual set of indicators would need to be developed in the context of a framework of agreed-upon aims for their use):

- **Input indicators.** Possible input indicators include:
    - Numbers of bottlers, distributors, retailers, recyclers, and others.
    - Quantities of materials used for containers (cans, plastic, glass refillables, glass non-refillables).
    - Energy required to produce various types of containers.
    - Number of private employees involved in different aspects of the program.

- Expenditures (capital and operating) associated with the program (for example, for equipment, fuel, and manpower).
  - Number of public employees involved in administering the program, including evaluation.
- Wages of state employees involved in the program.

- **Process indicators.** Examples include:
  - Expenditures (capital and operating) associated with the program (for example, on equipment, fuel, and manpower).
  - Duration of activities (such as collection, storage, and transportation).
  - Points of origin of deposits.
  - Number of complaints filed about the operation of the system.
  - Number of violations and fines.
  - Number of bottles returned.
  - Number of miles driven by container trucks.

- **Outcome indicators.** Among the possible indicators are:
  - Volume and type of litter.
  - Health effects.
  - Volume and type of solid waste.
  - Environmental impacts such as levels of atmospheric pollution, water resource pollution, and production of industrial solid wastes.
  - Employment gains and losses.
  - Beverage sales.
  - Interest earned on unclaimed deposits.

BOX 3-6
A COMPARISON OF BEVERAGE DEPOSIT LAWS

The provisions of the mandatory beverage deposit laws in the nine states that have them vary, as shown in the table below. Michigan is the only one with a deposit fee greater than 5 cents, and it alone has no control over deposits not refunded to consumers. In theory, Michigan's less conventional approach to deposit laws offers bottlers and distributors a greater incentive because it permits them to retain unclaimed deposits, a source of revenues that can offset the costs of the program. *continued on following page*

BOX 3-6 *continued from preceding page*

State (in order of effective date)	Amount of deposit	Handling fee	Administrative entity	Initiation level of deposits	Penalties	Special features
Oregon 1972	5 cents minimum; 2 cents for certified containers	No fee	Liquor Control Commission	Retailer to distributor	Suspension or revocation of license; Class B misdemeanor	Limits returnables to 96 per person per day
Vermont 1973	5 cents minimum	Greater of 2 cents or 20% of deposit	Department of Environmental Conservation	Consumer to retailer	Maximum fine of $1,000 for each violation	• Educational program • Bans sale of non-refillable glass containers
Maine 1978	5 cents minimum	2 cents	Department of Agriculture	Retailer to distributor	Maximum fine of $100	Limits returnables to 240 per person per day
Michigan 1978	10 cents minimum; 5 cents for certified containers	No fee	Liquor Control Commission	Retailer to distributor	Fines of $100-$1,000 per day of violation plus costs of prosecution	Not required to accept returnables for a refund in excess of $25 per day
Iowa 1979	5 cents minimum	1 cent	Department of Water, Air and Waste Management	Retailer to distributor	Simple misdemeanor	• Deposit required on liquor bottles • $100,000 allotted annually for treatment of alcoholics

Source: Karen L. Mallory and R. Atta Charo, "Federal and State Mandatory Beverage Container Legislation," *Columbia Journal of Environmental Law* 11 (355)(1986):377.

- New technologies.
- Beverage prices.
- Revenue from unclaimed deposits.

Performance indicators are used effectively in conjunction with judgments about the expectations and effectiveness of the law. Although no evaluation plan accompanied the 1976 Michigan law, indicators would help determine what the law has accomplished, what its effects have been and what opportunities and constraints relate to implementation. A strong system of performance indicators would also contribute to the adoption of market incentives in the environmental field by other states and localities.

CONCLUSIONS

The Michigan bottle bill is at the end of the continuum of market incentives where the government has a minimal administrative role and little responsibility. It demonstrates that the use of market instruments for environmental purposes in which the role of government is minimal is possible. In several other states with deposit bills, the administrative burden is somewhat greater, because they administer the revenues resulting from unclaimed deposits. (See box 3-6 for a comparison of the Michigan bottle law with those in other states.)

Four things are certain about Michigan's law:

(1) According to interviewees, the citizens of Michigan are pleased with how it operates. Recent data are not available, however. A January 1980 survey by the Market Opinion Research Co. asked the Michigan public, "If you had the opportunity to vote again on the issue of mandatory deposits on bottles and cans, how would you vote?" Seventy-eight percent of the respondents said they would vote yes. An opinion poll conducted by the *Grand Rapids Press* in December 1978 asked randomly selected citizens of that city, "Do you consider the new law on returnable bottles and cans a nuisance?" More than 67 percent of respondents stated no.

(2) Implementation of the Michigan deposit program has not been evolutionary. That is, amendments to the law have been minor, and no mid-course corrections of operations have been undertaken. Other states have also experienced difficulties with respect to changes in their laws. For example, in 1993 Connecticut Governor Lowell P. Weicker introduced a bill to repeal the state's deposit law and replace it with a non-refundable 5-cent tax. The proposal has created a great deal of controversy, with environmentalists, recyclers, farmers, bicyclists, industrialists, legislators, and the public at large all opposing it. Even though the tax is expected to raise $63 million to be earmarked for environmental programs and parks, the Connecticut Conference of

Municipalities is against the proposal because of potential increases in the costs of recycling and waste collection. Some also voice concern about the effect the repeal might have on the homeless, many of whom depend on the money they earn from collecting cans and bottles.

(3) The costs of implementing the Michigan system are high for the private sector. Whether they are offset by unclaimed deposits is uncertain.

(4) The costs have been unevenly borne by bottlers, distributors, and retailers. This situation raises another issue—could the money spent on the law have been put to better use on behalf of litter control and solid waste management?

As successful as Michigan's program has been in terms of the return rate—an estimated 93-95 percent—and despite a public that strongly supports the deposit system, it is does not seem to be replicable. As noted, no state has succeeded in getting a deposit law passed since 1983. Similarly, national legislation has not been passed. A comprehensive analysis of why deposit legislation has not made progress would need to include an examination of many monetary and non-monetary costs and benefits. Needless to say, other states and countries have to consider costs and benefits before looking at the practicality of beverage deposit-refund systems and similar systems for other products, such as batteries and tires.

Finally, it should be noted that proper and periodic assessment of Michigan's program through data collection, monitoring, and evaluation could be used to provide guidance on whether the law should be modified. For example, rigorous and timely data on the costs and benefits of the program could help in deciding whether to alter the way unclaimed deposits are handled or how to overcome the constraints affecting the planning and implementation of curbside collection programs.

4
APPLICATIONS OF ECONOMIC INSTRUMENTS

As the case studies make clear, economic instruments may be an alternative or supplement to conventional command-and-control-based environmental regulation. Further, economic instruments usually exist within a fabric of command-and-control regulations—in fact, some regulations are hybrids of these two approaches. Economic instruments are not, however, foolproof or panaceas, nor are they appropriate for all environmental problems.

The challenge for legislators and policy-makers is to know how to use economic instruments, to recognize when they are a better means than command-and-control regulation to achieve certain environmental goals, and to learn how to blend the two models to achieve regulatory goals. Where legislators and policy-makers determine that economic instruments are better, how do they decide which instrument is best-suited to a particular problem? Having decided a particular economic instrument is the right one, how can they determine whether the administrative systems, skills and capabilities, organizational culture and other factors needed to implement it are present, or can be created?

The case studies examined four tools—trading of emission reductions, pollution charges, variable residential trash charges and deposit-refund programs. The firsthand experience brought to light in the case studies yields insights into how and when trading and charge instruments can be applied to an environmental problem. The key findings derived from the case studies are summarized in box 4-1 and then are discussed in greater detail below. Note that the findings related to the economic instruments themselves are divided into: "threshold" factors—factors that help determine whether an economic instrument is appropriate for a given problem; and "design" factors—characteristics that should be built into the design of the actual application of an instrument for it to succeed.

TRADING OF EMISSIONS REDUCTIONS

If properly designed, trading programs use the dynamics of market forces to allocate the costs of preventing and reducing pollution among polluters in a more economically efficient way than command-and-control instruments do.

BOX 4-1
KEY FINDINGS FROM THE CASE STUDIES
TRADING OF EMISSIONS REDUCTIONS

Threshold Factors

- Trading works best with problems that manifest themselves over a large geographical area.
- Trading is likelier to work with a large number of participants.
- Trading must have substantial benefits over the less flexible status quo.
- Pollution sources must have available a range of pollution control techniques to choose from or have different marginal costs of control.
- Trading programs require sophisticated measurements because trading may establish quasi-property rights, but that capability may strengthen enforceability.
- When making pollution control decisions, businesses demand certainty about the nature of the commodity that is being traded.

Design Factors

- The limits on emissions established under command-and-control programs are often an inappropriate baseline for emissions trading.
- The governing agency must be proactive.
- The economic instrument should distribute costs and obligations evenly.
- The system must be flexible.
- Significant penalties must be established for non-compliance.

POLLUTION CHARGES: RESIDENTIAL TRASH CHARGES, EMISSIONS CHARGES, AND DEPOSIT-RETURN PROGRAMS

General Threshold Factors

- The uncertainty of the environmental outcomes of pollution charges should be carefully considered.
- For a pollution charge to be and remain an economic incentive, it must be targeted at sensitive points in the pollution system and set at cost levels high enough to constrain the pollution.

Volume-Based Charge for Trash Collection

Threshold Factors

- A volume-based charge is not a sufficient economic incentive for all polluters.

BOX 4-1

- Pollution charges function as a symbolic "moral" incentive for some polluters even though the charge is too small to affect economically motivated behavior.
- Pollution charges can have unintended, counterproductive effects.

Design Factors

- The links between the charges and the curbing of pollution should be clear.
- Revenues from charges and environmental protection activities should be linked.

Container Deposit-Return Programs

Threshold Factors

- Expansion of charge systems to include a wider range of products requires careful attention.
- An economically viable use for returned items is essential.
- Someone must benefit economically from receiving and handling the deposit-return items.

Design Factors

- Returning the item must be convenient for the consumer of the product.
- In crafting a refund system, the creation of windfalls should be avoided.
- Deposit-refund systems are self-implementing, and, once implemented, there is little reason to go back and modify provisions. Because subsequent adjustments to programs are rare, they must be done "right" the first time.
- Program adjustments are difficult once the program has been implemented.

Nationally Imposed Charges on Industrial Discharges

Threshold Factors

- The charges must relate to the firm's and regulator's ability to monitor discharges.
- The charge systems must relate to the permit and enforcement systems.
- Liability rules must be established.

Design Factors

- Involvement of multiple jurisdictions in designing the system may weaken administration of the emissions charges.
- Pollution charges should be harmonized with other taxes and subsidies.

BOX 4-1

INSTITUTIONAL CAPABILITY TO IMPLEMENT AND OPERATE ECONOMIC INSTRUMENTS

- Economic instruments and command-and-control programs require different organizational staffs and cultures.
- Different economic instruments vary in the degree of involvement by the governing agency.
- Increasing complexity and rising charge levels may strain the institutional capability of the governing agency to impose and collect the revenues.
- Generally, economic instruments change the role but do not eliminate the need for a responsible managing agency.
- The distributive effects of economic instruments should be carefully assessed.

SYSTEMATIC PROGRAM EVALUATION

The Roles of Evaluation

- Evaluation is essential to prevent the degradation of economic incentive programs.
- Evaluation can contribute significantly to the success of existing economic incentive programs and to their wider adoption.
- A strong evaluation component, well-integrated into the program, can help assure stakeholders the program will be implemented and managed so as to attain its promised results.
- Evaluation greatly enhances implementation of economic instruments in command-and-control organizational cultures.
- Evaluation provides data essential to changing the knowledge, attitudes, and behavior of those implementing economic instruments.

Design Considerations in Implementing Evaluation Programs

- Evaluation programs are strongest when they stipulate in advance clearcut objectives, responsible activities, and measures for their implementation and their expected effects.
- Evaluation needs to focus on results.
- An effective evaluation program is concurrent rather than projective, and utilitarian rather than theoretical.
- Evaluation must be systematic and continual in the program, not ad hoc.
- The formulation and collection of evaluation data need to be actively managed.
- Creating empirical data and providing the data electronically to the public are fundamental to evaluating economic instruments.

> **BOX 4-1**
>
> - In designing evaluation of the administration/implementation of economic instruments, it is particularly important to understand and address the expectations of diverse stakeholders.
> - Individual perceptions are a key factor in the emergence of markets that serve public goals and must be assessed regularly during implementation.
> - As measures of outcomes are formulated, stakeholders and others must be educated about the timeframe so that they do not become skeptical when results do not emerge instantly.
> - Evaluation must be tailored to the specific conditions and expectations in the setting in which the economic instrument operates.
>
> **Institutional Considerations in Implementing Evaluation Programs**
>
> - Institutional support for evaluation will need to be developed.
> - Economic instruments must be actively managed to be sustained successfully in the face of dynamic economic conditions.
> - Evaluation of incentives should be accompanied by a broad understanding of the political context in which economic incentive programs are enacted or changed.

Trading programs, or marketable permits, offer more options to industry and foster greater innovation than do traditional prescriptive laws and regulations. For trading systems to work, several institutional factors such as the capability of the governing agency to design and implement the programs must be addressed.

Following are the key findings relating to the implementation of trading programs. They are derived from the case study on the SCAQMD and its experience with offset trading and the development of RECLAIM.

Threshold Factors

Trading works best with problems that manifest themselves over a large geographical area.

Trading systems are most applicable where the pollutants affect a large geographical area or where several firms impact upon the same problem. Trading does not work well where there are hot spots or acute problems. Examples of pollutants are sulfur oxides, nitrogen oxides, and volatile organic compounds in Southern California. That does not mean to say the *pollutants* must be homogeneous. Different gases have different atmospheric life cycles and radiation-forcing potentials. Trading instruments can be effectively applied to a homogeneous problem caused by heterogeneous pollutants when information is available to permit the calculation and application of correct inter-pollutant ratios. For example, ratios can be applied to pollutants such as sulfur oxide and nitrogen oxide, because there is a high degree of certainty about the atmospheric life cycles of each and their effects on atmospheric chemistry. In cases where different pollutants are cross-traded, establishing and assuring the correct ratios for trades are crucial to having an environmentally comprehensive program. This process may be complex. Thus, policymakers will be faced with determining whether the cost-savings outweigh the complexities in designing the program.

In contrast, establishing ratios for greenhouse gases is problematic. For example, carbon dioxide is emitted through deforestation and mobile sources that burn fossil fuels. Methane is emitted through a range of applications, from coal mining operations to natural gas production and through the growth of crops such as rice. Halocarbons are produced to a great extent by industry. All these gases have greatly different atmospheric life cycles and radiation-forcing potentials. Because information is not readily available about their impact, and given that monitoring their non-industrial sources is impossible, implementing a trading scheme that encompasses many types of gas may be difficult.

Likewise, trading toxic emissions based on some kind of index of acute or chronic toxicity is difficult, if not impossible, to accomplish in the real world. Further, when materials that may be considered toxic are traded for other reasons, some type of toxic review or program to ensure toxic hot spots do not develop is required.

Trading is likelier to work with a large number of participants.

Trading instruments best fit environmental problems that involve a large number of potential participants. Under the acid rain trading program and RECLAIM, the number of potential traders is in the hundreds. Put another way, the potential to develop a market with a sizable number of traders must exist—the trading area must have a sufficient number of firms so that a demand for trade exists. For example, limiting trades to new or modified sources (offsetting) has led to sporadic markets for emissions reduction credits. The markets give potential generators of ERCs confidence that if they can generate ERCs, there will be a demand for them. However, including existing sources that must meet control requirements will considerably increase demand, and that increased demand will boost the supply of ERCs being created. Thus, not only does industry benefit because low-cost compliance opportunities can be created, but the environment also benefits when the ERCs are kept in emissions banks instead of the air. When trade-off ratios are set at greater than 1:1, the environment benefits.

The above said, it may be possible to set up a trading system in a small geographical area with only a few sources. In such a setting, the trading program can be based on *internal* trading among firms, using a form of the bubble instrument, instead of trading emissions reductions through an *exter-*

nal market using a form of marketable permits, as in the Los Angeles area. The bubble approach, which is normally employed in the location of hot spots, does not, however, embody the notion of perfect competition for emissions rights because, if restricted to only within-plant trades, there is no competition at all across emitting sources. Although environmental and economic benefits will accrue, the results will be more modest than under a more robust trading system.

Trading must have substantial benefits over the less flexible status quo.

Acceptance of emissions trading or pollution charge programs by stakeholders, including industry, environmentalists, and public officials, is essential to their full implementation. Critical to that acceptance is that stakeholders perceive command-and-control instruments as less cost-effective than trading or pollution charge instruments. If potential buyers perceive the costs of searching for marketable permit traders to be high, if the costs associated with negotiation of quantifying expected reductions is difficult, or if administration of the monitoring and enforcement requirements exceeds the costs of conventional command-and-control instruments, acceptance will not be forthcoming. Over time, many firms have grown comfortable with the existing, albeit relatively expensive, system, and many have said (in so many words) that they prefer the devil they know to the devil they don't know. The result is that some firms may fight to preserve the status quo because of the uncertainty of the transaction costs associated with the new program.

Pollution sources must have available a range of pollution control techniques to choose from or have different marginal costs of control.

The opportunity to sell at a profit and buy at cost-savings and to create and innovate is at the heart of the trading approach. Therefore, trading of emission reductions requires a supply of relatively inexpensive emissions reduction control techniques from which some polluters can choose to achieve compliance. If there is little or no choice of control techniques, the cost-savings will be insufficient to generate a supply of emissions reductions that can be sold at a profit or bought at a cost-savings, and creativity and innovation will not come into play. Absent a variety of available control technologies, marketable permit systems become merely quotas that are eventually allocated to their best economic use in the same manner as liquor licenses.

Trading programs require sophisticated measurements because trading may establish quasi-property rights, but that capability may strengthen enforceability.

A fundamental requirement of market operations is a strong capacity to measure emissions reductions on a continual basis precisely and accurately. The sampling and recording system may cost more than measurement does under command-and-control instruments. However, a strong measurement capacity enhances the opportunities for accountability and for enforcement, a factor that may have resulted in the political acceptability of the program in the first place.

If an infrastructure for measuring emissions is not already in place, it is important to consider carefully the cost of creating it, so that emissions reductions do not become too expensive to create, sell, and buy. In areas where the pollution problem is severe, the configuration of the system may already be in place, and the cost of additional improvements to the tracking system, particularly relative to the cost of command-and-control instruments, may be low.

When making pollution control decisions, businesses demand certainty about the nature of the commodity that is being traded.

The design and administration of trading instruments must be predictable enough that prospective sellers and buyers of emissions reductions feel comfortable making business decisions. This is particularly true in an environment where a plenitude of other factors compounds the complexity of decision-making. Despite their disadvantages, command-and-control instruments offer a level of predictability that has economic value to the regulated community. Uncertainty will affect the current and future values of allowances or emissions credits, so that the choices buyers make may be distorted.

As a corollary, firms must understand the nature of the commodity that is being created. Uncertainty about the use and duration of the emissions reduction, allowance, or marketable permit will significantly diminish the creation and use of these instruments.

Design Factors

The limits on emissions established under command-and-control programs are often an inappropriate baseline for emissions trading.

One of the principal tasks the agency implementing a trading program has is to set the baseline limits against which pollution sources gauge the amount of emissions reductions they can sell or need to buy. These limits should be set at levels sufficient to meet or maintain ambient environmental standards and that should provide an incentive to reduce compliance costs. They should also be set at a level that ensures that firms do not exceed the emissions limits or trade more emissions than they have lessened. Fundamental to trading programs is that various stakeholders help determine baseline levels.

Typically, emissions trading programs such as RECLAIM are applied to problems that are already subject to command-and-control regulation.[1] In

---

[1] Trading programs can also be used in situations where existing command-and-control systems do not exist. As such, they can be entirely new initiatives, such as the proposed international carbon trading programs.

many of these cases, baseline limits will already have been established. Therefore, dovetailing the existing program to a new regulatory program that monetizes entitlements can create considerable design problems. When designing the trading program, it is important to recognize early on that command-and-control-based baselines will not automatically be the baselines against which sources will create tradable emissions reductions. Doing so could have undesirable political, economic, and environmental consequences. The reason is that under command-and-control regulation, a key criterion in setting the limits is that they be affordable to sources in weak financial conditions. As such, they may not have been set at a level that will achieve an environmental goal in the regulated area. That is, under a command-and-control program, the linkage between the emissions limits and ambient environmental goals is often weak.

When command-and-control baselines rather than market-determined ones are used in emissions trading programs, they can deter trading. The reason is that they tend to create difficulties for new firms, which must reduce emissions by installing controls or must offset emissions through reductions. Unless the baselines are carefully fixed, they tend to favor firms that have delayed or opposed pollution controls. To redress the bias against new firms, regulators could require increases in emissions reductions by existing firms and set aside some of those decreases, which are referred to as growth allowances or community banks, for sale to new firms.

The governing agency must be proactive.

The governing agency should not be locked into making marginal or incremental improvements. It should be willing to invest considerable resources in an ambitious and aggressive manner. Market incentive programs tend to reverse assumptions inherent in traditional command-and-control-style regulatory programs and processes. For example, under a market incentive system, it may be assumed that the marketplace will generate and disseminate information concerning technology and that the responsible managing agency will facilitate these processes.

The economic instrument should distribute costs and obligations evenly.

While economic instruments are usually sought out because of economic efficiency properties, regulatory designers cannot be blind to political realities. The design of the system should consider a fair balance of costs and obligations that poor and rich industries will have to bear so that these programs are politically acceptable. Moreover, getting a variety of potential participants to enter the market is good for all, because frequent trades create stable prices and less uncertainty. In the case of RECLAIM, the equity of allocations, including who is responsible for past and current aspects of the problem, became a central focus of the negotiations with stakeholders. The roles and responsibilities of small businesses featured prominently in the discussions of key decision-makers. In addition, in the case of sulfur dioxide allowance trading, Congress made several adjustments to redistribute costs and obligations among electric utilities.

The system must be flexible.

The system of allocation and reallocation should be flexible enough to accommodate changes in targets and to add new sources, if possible. Revisions of the system, which may include the reissue of permits on an overlapping basis, should not result in perceptions of instability. While this may appear to conflict with businesses' need for certainty, as long as they are able to predict mid-course corrections, they can make adjustments to their compliance strategies. Evaluation systems are mechanisms for predicting mid-course modifications.

Significant penalties must be established for non-compliance.

A tradable system will not work unless the system monitors compliance and the monetary forms of sanctions are significant enough to discourage non-compliance. They must also be imposed.

POLLUTION CHARGES: RESIDENTIAL TRASH CHARGES, EMISSIONS CHARGES, AND DEPOSIT-RETURN PROGRAMS

This study examined three pollution charge instruments: charges for the collection of residential trash, the aim of which is to minimize household waste and foster recycling; deposits on beverage containers, whose aim is to minimize litter; and nationwide charges on emitters of an extensive list of chemical pollutants. In each case, the instrument was designed to serve as an economic incentive, rather than as a revenue-raising measure.

Following are the key findings to emerge from the three case studies of pollution charges.

General Threshold Factors

The uncertainty of the environmental outcomes of pollution charges should be carefully considered.

The environmental outcomes that can be expected from pollution charges may be highly uncertain because frequently the charges have no closed, direct connection with outcomes. The actual operation of even the most carefully formulated pollution charge instruments can be dramatically affected by macroeconomic forces, the importance of energy-intensive industries to certain areas, and political considerations. Product charges, in particular, are dependent on the price elasticity of the demand for products and the availability of substitutes.

For a pollution charge to be and remain an economic incentive, it must be targeted at sensitive points in the pollution system and set at cost levels high enough to constrain the pollution.

Many of the pollution charges introduced as economic incentives to prevent or control pollution get transformed into means to raise revenue. When this occurs, they may achieve modifications in behavior, but they do not operate as true incentives. The reason is that almost all revenue-raising charges are levied on low-visibility, low-sensitivity points in a pollution-generating chain, at a rate too low to influence the behavior that leads to the pollution. Even though the revenues raised are put to environmental use—for example, to finance the operation of command-and-control regulatory agencies, fund research, or pay for cleanups—polluters are not charged in a way that induces them to change their polluting behavior. As the Russian case study illustrates, charges that were initially introduced have become insignificant over time because of inflation. On the other hand, charges may be initially set at a politically acceptable level and increase over time.

Volume-Based Charge for Trash Collection

Threshold Factors

A volume-based charge is not a sufficient economic incentive for all polluters.

The amount of the charge will be an economic incentive to some, not all, polluters. One reason is that the charges cannot be set high enough to affect all members of the polluting population. In practice, then, in many settings pollution charges have to be supplemented with other instruments—such as publicity and education—aimed at influencing the behavior of the people or organizations that are insensitive to charges set at viable economic and political levels. Pollution charges, especially when applied to a general population such as all residents of a county, must (1) be set at an optimum, rather than maximum, level and (2) be reinforced or supplemented by some other instrument that serves as an incentive to change the environmental behavior of those who do not respond to the economic incentive.

Pollution charges function as a symbolic "moral" incentive for some polluters even though the charge is too small to affect economically motivated behavior.

Among populations where there is a relatively high and widespread level of concern for environmental protection, or such concern can be developed, volume-based charges can operate as a powerful moral incentive to constrain polluting behavior. Polluters who are insensitive to the level at which a charge is set may be sensitized by educational outreach efforts to view the charge as a measure of their morality. In King County, the symbolic moral halo effect of pollution charges requires that the pollution charge program include public education to reinforce this linkage where it already exists and to create it where it does not.

To get this moral effect as well as the economic effect of volume-based charges on polluters' behavior, the amount of pollution and the charge must be explicitly and visibly linked. In addition, other non-economic forces such as education may be used to achieve this effect. Further, the revenues from the pollution charges can be applied to environmental services that prevent or clean up pollution. It should be noted that in certain cases an attitude on the part of some polluters that "it's okay—I paid for the right to pollute" may negate the moral effect.

Pollution charges can have unintended, counterproductive effects.

Legislators and policy-makers should be alert for factors in the design, administration, and oversight of pollution charge instruments that could lead to unintended, counterproductive outcomes. For example, assume the charge program is funded primarily by the revenue from the pollution charge. If the program is successful, the revenue is likely to decline as the volume of pollution decreases. Mid-course adjustments in the level of the charge to sustain the level of revenues needed to operate the program may result in a lack of support for the program.

Design Factors

The links between the charges and the curbing of pollution should be clear.

Residential waste pick-up charges are near universal. However, relatively few link the amount of the charge to the amount of waste being disposed. Often the cost of waste collection/disposal is embedded in other charges, such as property taxes and water/sewer charges. When the costs are embedded, residents cannot reduce them by lessening the amount of household solid waste and therefore have no incentive to cut back the volume of their trash. For pollution charges to function as economic incentives, there must be an explicit and visible link between the amount polluters are charged and the amount of pollution they dispose of.

Revenues from charges and environmental protection activities should be linked.

Closely linking the application of the charge revenues and environmental protection activities pays off in two ways: (1) it provides practical financing for needed programs and (2) it reinforces the political linkage between the charge and public concern that the jurisdiction address environmental problems.

Container Deposit-Return Programs

Threshold Factors

Expansion of charge systems to include a wider range of products requires careful attention.

The basic model of the beverage container deposit-return program studied here may be adaptable to other disposable items—for example, used tires and batteries. Such adaptation will, however, require careful adjustments in the program's design to address product-specific and locale-specific economic dynamics.

An economically viable use for returned items is essential.

Operation of deposit-return instruments requires that there be some economically viable use for the returned items. In addition, for a deposit-return instrument to function *as an economic incentive*, the value of that use must be greater than the cost of returning the item plus the handling costs. Even the simplest application of deposit-return instruments—refillable beverage bottles—requires careful consideration of situation-specific logistics and associated handling costs, lest the costs exceed the value of the deposits.

Someone must benefit economically from receiving and handling the deposit-return items.

Deposit-return instruments require some operation to receive and temporarily hold returned items. Typically this role is assigned to retailers. Product retailers may oppose deposit-return programs because of the burden and handling cost of temporarily storing the returned items. As shown in the Michigan case study, retailers' criticism of the program may be sustained over decades. Producers or wholesalers could be required to set up facilities to receive the products or to finance third-party receiving and handling operations. For example, legislation might authorize consortia of producers to establish systems for receiving returned items. The initial political viability of a deposit-return instrument may depend heavily on either (1) providing sufficient incentive to product-receiving retailers to take that role or (2) assigning the role of receiving returned items to some player other than the retailer.

Design Factors

Returning the item must be convenient for the consumer of the product.

The convenience of return may be a more important consideration in designing a deposit-return instrument than the actual amount of the deposit; desired behaviors are highly sensitive to convenience.

In crafting a refund system, the creation of windfalls should be avoided.

In crafting a deposit-return program, it is very important to consider how the revenue from product deposits is to be allocated. There are four candidate recipients: (1) the business that generated the item, such as the bottlers of soft drinks; (2) the purchaser/consumer of the product; (3) the retailer (or whoever) is designated to receive the returned item; and (4) whatever (if any) governing agency is responsible for administering or enforcing the program. Legislators and policy-makers should take care not to inadvertently create windfalls for any of the above players. For example, bottlers benefit more from the retained deposits on *unreturned* items than from returns. Such a windfall may compromise the incentive effects, undercut the effectiveness of deposit-return instruments to prevent improper disposal, and destroy the political support needed to adopt or sustain the program.

Program adjustments are difficult once the program has been implemented.

States that have attempted to modify provisions in deposit-refund legislation have experienced considerable difficulties because of the opposition of vested interests and a reluctance to change the status quo.

Nationally Imposed Charges on Industrial Discharges

Implementation of emissions charge instruments is a complex undertaking. Although the national emissions charge program studied here is Russia's, the political and organizational dynamics that make its implementation complex are not unique. The Russian experience with pollution charges is instructive for understanding the realistic operation of a national pollution charge-type instrument.

Threshold Factors

The charges must relate to the firm's and regulator's ability to monitor discharges.

Firms and regulators must monitor discharges. Absent accepted protocols and technologies for tracking emissions, effluents, and solid waste, the charge system is extremely difficult to monitor, penalties are difficult to assess, and enforcement may be too expensive. The charge systems must relate to the permit and enforcement systems. Under Title IV of the 1990 Clean Air Act, a continuous emissions monitoring system has been required for all firms covered by this law. CEMs provide the basis for either an enforceable marketable permit trading system or a pollution charge system. As noted, the Russian pollution charge system has two goals: controlling pollution and generating revenues for the environmental funds. Because of the evolving permit system and unclear monitoring requirements, it is uncertain whether discharges have been reduced as much as originally predicted. Absent good monitoring protocols and data, too many uncertainties will erode the foundation of a pollution charge system. After all, without a good monitoring system, regulators might conclude that a firm that was actually complying was out of compliance, or vice-versa. Neither error is good for building or maintaining the integrity of the pollution control program.

The charge systems must relate to permit and enforcement systems.

Permit systems, monitoring, and enforcement are all part of good environmental management systems. Permits are the basic tool for tracking environmental obligations and specifying enforceable discharge limits. While a pollution charge approach can be predicated on few if any technology requirements, such an approach must have, as a compensating mechanism, stringent permit, monitoring, and enforcement systems. This means that the administrative and legal structure for managing non-compliance must be certain.

Liability rules must be established.

Criminal and civil liability for non-compliance is the underpinning of any enforcement program. Authority can be delegated, but responsibility cannot be. Without clear liability rules, regulators are left with little that assures the public that regulated entities will follow the regulatory requirements. The Russian case study enlightens readers along these lines.

The above three threshold rules reinforce the idea presented throughout this report that economic incentive-based systems do not work in a vacuum and that the real-world context in which they work must include elements of a command-and-control-like system of compliance management.

Design Factors

Involvement of multiple jurisdictions in designing the system may weaken administration of the emissions charges.

As the scope of an emissions charge instrument broadens across a range of substances and governing entities, variations in administration—for example, in individual assessments or enforcement actions—tend to compromise the design of the program. Such compromises arise from the existence of local discretion, which may be politically necessary. The exercise of local administrative discretion tends to reduce both (1) the economic incentive and environmental effects of the charge and (2) actual revenues to a level below what was projected.

Pollution charges should be harmonized with other taxes and subsidies.

Emissions charges may be extremely difficult to harmonize with the operation of programs that conflict with the goals of the charge system. For example, taxes on methane emissions would be largely negated if farmers received subsidies to offset the cost of the taxes.

INSTITUTIONAL CAPABILITY TO IMPLEMENT AND OPERATE ECONOMIC INSTRUMENTS

Government agencies' operational role in command-and-control programs directed at environmental problems is quite different from their role in market/incentive programs. Similarly, government's role in emissions trading differs from its role in pollution charge programs. Legislators and policy-makers considering economic instruments should ensure that the implementing agencies are organized and staffed to play their new role. Generally, they do not pay attention to the connection between legislative mandates and the requisite resources to implement programs. In an era of downsizing government, this concern needs to be underscored.

Economic instruments and command-and-control programs require different organizational staffs and cultures.

The institutional design, staffing, and culture needed to implement and administer economic instruments differ significantly from that of command-and-control regulation. As a result, where environmental problems are currently or traditionally subject to command-and-control regulations, it may be necessary to bring in outside expertise to design the economic instrument. The reason is that existing agency personnel and constituencies may support the status quo and might try to stall, starve, or misinterpret an economic incentive program or translate it into a command-and-control operation under the mantle of market/incentive rhetoric.

Where the environmental problem is not already subject to command-and-control regulation, legislators and policy-makers must understand that the staffing and structure of the implementing agency are likely to differ significantly from those of agencies operating command-and-control instruments. Staff with a disciplinary mix ranging from psychology to economics and finance may be appropriate for the successful implementation of some programs. Through oversight and management, legislators and policy-makers should take care that the staff assembled and work done by the agency closely parallel the design of the economic instrument, so that it does not get translated in ways that are tantamount to command-and-control regulatory operations.

Different economic instruments vary in the degree of involvement by the governing agency.

The degree to which the governing agency is involved is different in each of the four economic instruments studied here. In one of the cases—the Michigan deposit-return program—the designers consciously minimized the role of government, which has almost no involvement. The other three instruments studied—Russia's pollution charges, Washington state's volume-based disposal charges, and the California South Coast emissions reduction trading—all require that governing agencies perform functions crucial to the operation of the instrument, and the agencies have been staffed accordingly. It should be fairly obvious that if a single agency is using several instruments, its degree of involvement may vary instrument by instrument.

Increasing complexity and rising charge levels may strain the institutional capability of the governing agency to impose and collect the revenues.

As the complexity and amount of the pollution charge increase, the institutional capability to enforce collection becomes increasingly important. Efficient and effective assessment and collection of the charges require special organizational staffing, operations, authority, and information-management capabilities. This kind of institutional capability is relatively limited in most environmental agencies. Typically, staff have been involved in detecting violations and assessing penalties on a relatively small scale. Further, the requisite training and experience of staff differ from that required of staff in agencies that administer the taxation of corporate income or tobacco and alcohol, whose purpose is primarily to raise revenues.

Generally, economic instruments change the role but do not eliminate the need for a responsible managing agency.

Legislators and policy-makers should not pursue economic alternatives to command-and-control regulation in the false hope that associated government agencies can be dismantled. The need for a governing infrastructure for economic instruments remains and may even grow. At the same time, the work of the governing agency will change, and its staff and structure will, as noted, have to change accordingly. Ensuring that appropriate agency structures and staffing for economic instruments actually emerge at the operating level may require special attention in legislative language, oversight, and management so that command-and-control requirements do not overshadow the economic operation or thwart the ability of the incentive to meet its environmental objectives.

The distributive effects of economic instruments should be carefully assessed.

The financial consequences of economic instruments vary among target groups. At the micro level, for example, small firms may face considerable short-term expenditures when economic instruments are introduced. If the consequences are considered, measures may be taken to correct for inequities. The consequences, it should be noted, may also relate to the redistribution of power within agencies and among stakeholders.

SYSTEMATIC PROGRAM EVALUATION

The long-term effectiveness of economic instruments and their widespread adoption hinge on integrating evaluation into the design and operation of economic incentive programs. Currently, the majority of economic instruments are administered without systematic measurement of their performance, either through self-examination or external oversight. Nor are the political effects of instruments taken into account. For example, how do the goals of programs change? How do the programs alter interactions between regulators and regulated interest groups? What happens to the public's conception of what it means to pollute? What impacts will the adoption of economic instruments have on electoral politics and future policies?

Management decisions are made primarily on intuitive and political grounds, sometimes supported by ad hoc analyses cobbled from convenient data and anecdotes. A major conclusion to emerge from this study is the importance of making systematic evaluation an integral component of the design of any economic instrument. Other conclusions related to evaluation are broken out here in terms of the roles of evaluation, design considerations, and institutional considerations.

The Roles of Evaluation

Evaluation is essential to prevent the degradation of economic incentive programs.

Absent systematic empirical evaluation, economic instruments may tend to stress traditional ways of doing things at the expense of revisions or to become transformed in adverse ways. For example, once they are established, they may become frozen and not keep up with changing conditions, or they may evolve into a set of administrative actions that adhere closely to the canons of command-and-control regulations. To recapitulate, evaluation can help program maintenance by guiding operations from both administrative and practical points of view.

Evaluation can contribute significantly to the success of existing economic incentive programs and to their wider adoption.

Lack of knowledge may hamper wider adoption of economic incentive programs. Legislators, policy-makers, stakeholders, and the public at large may be wary of using economic instruments for which there are only undocumented theoretical linkages to environmental results. Their wariness is especially great if the instrument increases citizens' tax burden. Evaluation can document the use of economic instruments, thereby contributing to decision-making about adoption. It can also point out discrepancies between ideal and actual performances, adding weight to the credibility of the instruments.

A strong evaluation component, well-integrated into the program, can help assure stakeholders the program will be implemented and managed to attain its promised results.

Typically there is initial skepticism about economic incentive programs. One reason for skepticism on the part of some stakeholders is that economic instruments place the accomplishment of environmental objectives in the hands of the forces and institutions that they believe are the *cause* of the problem.

Evaluation counters this skepticism by empirically and objectively measuring the attainment of goals and by fostering results-oriented administration in contrast to a process-oriented administration. It can play a crucial role in the attainment of public goals and in the realization of the auxiliary benefits that economic instruments promise, such as less intrusiveness and greater efficiency. The foregoing description of what the evaluation process can do strongly underscores the close interrelationship between evaluation and program planning and operation.

Evaluation greatly enhances implementation of economic instruments in command-and-control organizational cultures.

Economic programs superimposed on an older, established agency using command-and-control regulation risk being undermined. The governance strategy of economic instruments is so fundamentally different from that of command-and-control that the former can easily fall prey to formally stated, ritualistic processes and practices that mirror business as usual. Evaluation can offer public information on the extent to which the regulatory agency is doing what it is supposed to by unlocking critical insights. While the program is under way, continuing dialogue with administrators, practitioners, public officials, and other interested parties will help ensure that the program is consistent with market-based practices.

Evaluation provides data essential to changing the knowledge, attitudes, and behavior of those implementing economic instruments.

In the world of corporate stocks, in addition to *primary* data on such factors as the price of listed stocks and aggregate statistics about stock trading, secondary data on the products, management, business strategies, and financial condition of companies whose stocks are traded are also essential to fuel the trading of stocks. Such secondary subtle information could be necessary for market operations and the activation of certain incentives. In the world of economic instruments, analogous influential secondary information is necessary to spark market dynamics. In short, this information serves as a stimulus for action.

Design Considerations in Implementing Evaluation Programs

Evaluation programs are strongest when they stipulate in advance clearcut objectives, responsible activities, and measures for their implementation and their expected effects.

The classification of objectives, the development of criteria of success and failure, and the anticipation of results foster better evaluation by public and private players alike. They set the stage for rational debate and accommodation of stakeholder interests throughout the life of a program. They cut across all aspects of the program, from questions of funding to day-to-day operations to the winning of community acceptance.

Evaluation needs to focus on results.

To avoid having the evaluation focus on the whys and wherefores of implementation rather than on whether the program produces the desired outcomes, the degree to which goals are being met must be stressed. What did the program accomplish? Did changes occur? Were the changes the intended ones? What level of performance is judged to be adequate? What unintended impacts might the program have? These are some of the questions that guide a results-focused evaluation.

An effective evaluation program is concurrent rather than projective, and utilitarian rather than theoretical.

Typically there is little or no real-time or practical data on command-and-control programs. Nevertheless, comparisons with economic instruments are made. These comparisons are based on hypothetical extrapolations from the minimal empirical data collected before implementation of the regulatory programs. Typically the comparisons are based on projections from theoretical models stretched well beyond their empirically established predictive reliability. Sometimes the projections are patently ideological. Measurements of actual effects concurrent with or following actual implementation are not available. Evaluation of economic instruments must be designed to produce real-time, practical data.

Evaluation must be systematic and continual in the program, not ad hoc.

Measurement of the implementation, operation, and results of economic instruments requires systematic, ongoing evaluation. Without systematic and continual evaluation, the fundamental data needed to determine optimal combinations of program practices over time will not be available. In particular, the ability to redirect the program into more-productive channels can be enhanced through as rigorous an approach as possible within the constraints of the program setting.

The formulation and collection of evaluation data need to be actively managed.

The expectations of stakeholders about economic instruments change over time. Evaluations based on data drawn from "cold storage" data banks risk being dismissed as unresponsive to contemporary issues. At the same time, there should be a stable core of data that spans the time it takes to achieve the original goals of the program. Two important points follow. First, the volume of existing and new data must not be allowed to become overwhelming. Second, the logic for generating data must be well-understood.

Creating empirical data and providing the data electronically to the public are fundamental to evaluating economic instruments.

The first and most crucial step in evaluating economic instruments, given the availability of the technology, is to task government agencies with, and to support them in, generating program data and making the data available electronically to the general public for analysis. The EPA's Toxic Releases Inventory, mandated by Congress, is a harbinger of a fresh institutional approach. Anyone can download and analyze data from the TRI. This new role of the federal government (electronically providing data to the general public at no charge) and the proliferation of competent personal computer users significantly alter the balance of forces affecting this regulatory program.

There are several reasons the government agency should play a lead role in generating measurement data. First, it is uniquely positioned to generate data regarding inputs, processes, and outcomes related to the entities "regulated" by the program. Second, evaluation competes for agency funds and staff, without the support of a broad external constituency. Once government makes the data available, stakeholders with sufficient interest can be given the task of analyzing and interpreting the data for the public. Given that the cost of data analysis, interpretation, and reporting accounts for a substantial part of the expense of evaluation, making outsiders responsible for this task would free agency resources and create a greater incentive to formulate better measurements. Third, government agencies have long been generating and storing data for private study and analysis.

In designing evaluation of the administration/implementation of economic instruments, it is particularly important to understand and address the expectations of diverse stakeholders.

Many stakeholders are involved in the implementation of economic instruments. Each stakeholder embodies a certain set of assumptions about what the program should accomplish, what kinds of activities are appropriate, what sorts of results will be produced and how they will be communicated. This reality that implementation has different meanings to different stakeholders means the program evaluation must take into account multiple perspectives while emphasizing a perspective that most accurately assesses the program.

Individual perceptions are a key factor in the emergence of markets that serve public goals and must be assessed regularly during implementation.

An incentive stimulates action only if the people whose behavior the government wants to change perceive it to be an incentive. Economic instruments rely heavily on creating a psychological state that induces individual players to take actions that contribute to the goals of the program. The case of King County points to the presence of psychological incentives and their interactions with monetary incentives.

The players must also feel certain the incentives are fully present and sufficiently reliable. If implementation of an economic instrument by government falters and does not proceed on schedule, the players whose voluntary actions are the engine that energizes and delivers the program's results will grow skeptical and become reluctant to participate as desired.

Evaluation of programs should be designed, deployed, and managed so about challenge government to implement the program fully. Regular assessments of the perceptions of players about the economic instrument as it is introduced and subsequently matures should be a key element of the evaluation. Is there evidence of positive, supportive attitudes toward the program among those involved with its implementation?

As measures of outcomes are formulated, stakeholders and others must be educated about the timeframe so that they do not become skeptical when results do not emerge instantly.

Expectations and real-world data have a finite shelf-life, a point that is often overlooked in program evaluation. The results of economic instruments emerge over time, and the evaluation timeframe must reflect that dynamic. The point at which data are collected and at which the results are made available affect confidence in the evaluation. Too often well-designed programs flounder because expectations about them are too great or expectations regarding the amount of time it takes to make a difference are unrealistic.

To secure authorization to implement a program, proponents often promise overly ambitious results. Promises may also create expectations that results will happen sooner than is plausible. Managers and evaluators of economic instruments may find they need to educate stakeholders about the goodness of fit and the timing of results.

Evaluation must be tailored to the specific conditions and expectations in the setting in which the economic instrument operates.

As the discussion on evaluation presented with each case study illustrates, there is wide variation in the conditions and expectations the four economic instruments have had to deal with. The nascent RECLAIM emissions trading program offers a rare opportunity to find practical answers to fundamental questions about trading instruments, as well as to secure the fidelity of its operation to the concepts guiding RECLAIM.

The timing and conditions of implementation of the Russian Federation pollution charge program are very different from those of RECLAIM and require another approach to evaluation. Evaluation of Russia's program must be framed in a way that does not undermine the considerable achievement of its survival. At the same time, the evaluation must signal the crucial gaps between concept and reality in a way that stimulates the Federation to close the gaps.

In stark contrast to the Russian Federation's pollution charge program, Michigan's container deposit program enjoys tremendous stability. As such, it requires a very different approach to evaluation. Given the relatively low level of executive agency involvement in the program, it is especially important that the evaluation look at how such a "self-implementing" approach operates and how effective it is in meeting expectations.

The King County volume-based trash charge program operates in a proactive governmental milieu and has many small components that work together. That is, the volume-based charge is only one component of a larger waste management program that seems to be meeting many expectations. Evaluation of this one small, complex element of the whole must be carefully structured, lest the evaluation require more effort than the charge program itself. Nonetheless, opportunities—and need—for improvement will go unnoticed without evaluation.

Institutional Considerations in Implementing Evaluation Programs

Institutional support for evaluation will need to be developed.

Absent legal requirements, there is little or no reason for agencies to pursue evaluation of economic instruments. Unless administrators want information for specific reasons such as the support of new funding initiatives, the improvement of operations, or the enhancement of public relations, they might not conduct evaluations.

Among the institutional obstacles to program evaluation are several perceptions about evaluations: they are intrusive; they may be used as a tactic to delay needed actions; they may be used as a "gesture" intended to show objectivity; they may be used to destroy a program regardless of its worth; and they may be employed to disguise failure by shifting attention to just the positive aspects of the program.

The seeds for more-effective institutional support for evaluation can be found in Total Quality Management, which emphasizes empirical measurement of organizational processes and results. The proliferation of powerful personal computers and software, especially in concert with the increased availability of data in electronic format (for example, the TRI and the Clinton administration's notion of a "data superhighway"), holds promise for evaluating economic instruments. This technology has revolutionized the capability to analyze data and communicate the analyses to the public.

Economic instruments must be actively managed to be sustained successfully in the face of dynamic economic conditions.

Once an economic instrument is fully operating, the orientation of evaluation will need to change slightly. Consistent with the need to make mid-course corrections and fine-tune the instrument, the entity responsible for implementation will need to monitor the program routinely. Like a physician who, before making a final diagnosis, couples tests with observation and questioning, monitoring should occur to ensure the program is working as intended. Newspaper and plastic recycling programs have, for example, reported the emergence of counter incentives in their recycling programs as the market for newsprint or plastic becomes saturated or declines. Localized or industry-specific slumps in business can significantly affect patterns of implementation.

Evaluation of incentives should be accompanied by a broad understanding of the political context in which economic incentive programs are enacted or changed.

Stakeholders should have an understanding of legislative histories, committee reports on bills, floor debates and the like in order to be aware of the salient issues, trade-offs, and compromises involved in the introduction, adoption, and modification of incentive programs. For example, understanding the positions of Michigan's conservation groups and beverage industry is essential to learning why the deposit law has not been transformed. A constituency program model explains why unclaimed deposits remain in the hands of bottlers and why legislation to change this situation has resulted in opposition.

GLOSSARY

AQCR	Air Quality Control Region
AQMD	Air Quality Management District
AQMP	Air Quality Management Plan
ARB	Air Resources Board (California)
BACT	Best available control technology
Btu	British thermal unit
CAA	U.S. Clean Air Act
CARB	California Air Resources Board
CEMI	Central Economic and Mathematical Institute (USSR)
CEMS	Continuous emissions monitoring system
CERCLA	Comprehensive Environmental Response, Compensation and Liability Act
CO	Carbon monoxide
CPU	Central processing unit
CRS	Congressional Research Service, Library of Congress
EPA	U.S. Environmental Protection Agency
EPL	Environmental Protection Law
EPT	Equivalent pollution ton (Russia)
ERC	Emission reduction credit
ETPS	Emissions Trading Policy Statement
FTE	Full-time equivalent
GAO	U.S. General Accounting Office
GNP	Gross national product
LAER	Lowest achievable emission rate
MEP&NR (Russia)	Ministry of Environmental Protection and Natural Resources
MPC	Maximum permitted concentration (Russia)
MPD	Maximum permitted discharge (Russia)
NAAQS	National Ambient Air Quality Standard
NAPA	National Academy of Public Administration
NO_x	Nitrogen oxide
NRC	Natural Resources Commission (Michigan)
NRDC	Natural Resources Defense Council
NSPS	New Source Performance Standards
NSR	New source review
O_3	Ozone
OECD	Organisation for Economic and Co-operative Development
Pb	Lead
PCB	Polychlorinated biphenyl
PM10	Particulate
PPM	Parts-per-million

PSD	Prevention of significant deterioration
RACT	Reasonably available control technology
RECLAIM	Regional Clean Air Incentives Market
RFP	Reasonable further progress
RTC	Reclaim Tradable Credit
RTU	Remote terminal unit
SCAG	Southern California Association of Governments
SCAQMD	South Coast Air Quality Management District
SIP	State implementation plan
SO_2	Sulfur dioxide
SO_x	Sulfur oxide
TPC	Temporary permitted concentration (Russia)
TPD	Temporary permitted discharge (Russia)
TRI	Toxic Releases Inventory
VOC	Volatile organic compound
WMI	Waste Management, Inc.
WUTC	Washington Utilities and Transportation Commission

SELECTED BIBLIOGRAPHY

Ackerman, Bruce A., and William T. Hassler. "Beyond the New Deal: Reply." *Yale Law Journal* 90 (6)(May 1981):1412-34.

———. *Clean Coal/Dirty Air*. New Haven: Yale University Press, 1981.

Ackerman, Bruce A., and Richard B. Stewart. "Reforming Environmental Law." *Stanford Law Review* 37 (5)(May 1985):1333-65.

———. "Reforming Environmental Law: The Democratic Case for Market Incentives." *Columbia Journal of Environmental Law* 13 (1988):171-99.

Air Pollution Control Association. *Proceedings—Economic Incentives for Clean Air Specialty Conference*. Pittsburgh, Pa.: Air Pollution Control Association, 1981.

American Petroleum Institute. "Background Paper on the Use of Economic Incentives for Environmental Protection." American Petroleum Institute, Washington, D.C., 1980.

Anderson, Robert C., Lisa Hoffman and Michael Rusin. "The Use of Economic Incentive Mechanisms in Environmental Management." American Petroleum Institute Research Paper No. 51. American Petroleum Institute, Washington, D.C., 1990.

Atkinson, Scott, and Tom Tietenberg. "Market Failure in Incentive-Based Regulation: The Case of Emissions Trading." *Journal of Environmental Economics and Management* 21 (1)(July 1991):17-31.

Baumol, William J. "The Public-Good Attribute as Independent Justification for Subsidy." *Intermountain Economic Review* 8 (1977):1-10.

Baumol, William J., and Wallace E. Oates. *The Theory of Environmental Policy*. Englewood Cliffs, N.J.: Prentice-Hall, Inc., 1975.

———. *The Theory of Environmental Policy*. 2nd ed. Cambridge University Press, 1988.

Brady, Gordon L., and Blair T. Bower. "Effectiveness of the U.S. Regulatory Approach to Air Quality Management: Stationary Sources." *Policy Studies Journal* 11 (1)(September 1982):66-76.

Brady, Gordon L., and Richard Morrison. "Emissions Trading: An Overview of the EPA Policy Statement." National Science Foundation, Washington, D.C., 1982.

Breyer, Stephen. *Regulation and Its Reform*. Cambridge, Mass.: Harvard University Press, 1982.

Bromley, Daniel W. "Entitlements, Missing Markets, and Environmental Uncertainty." *Journal of Environmental Economics and Management* 17 (2)(September 1989):181-94.

Brown, Gardner M., Jr., and Ralph W. Johnson. "Pollution Control by Effluent Charges: It Works in the Federal Republic of Germany, Why Not in the U.S." *Natural Resources Journal* 24 (4)(October 1984):929-66.

"The Bubble Upheld." *Regulation* 8 (May-June 1984):5-7.

Buchanan, James M., and Gordon Tullock. "Polluters' Profits and Political Response: Direct Controls Versus Taxes." *American Economic Review* 65 (1)(March 1975):139-47.

Butler, Chad. "New Source Netting in Nonattainment Areas under the Clean Air Act." *Ecology Law Quarterly* 11 (3)(1984):343-72.

Caldwell, Sean M. "The Trouble with Bubbles: Ramifications of Agency Delay in Approving SIP Revisions." *Northern Kentucky Law Review* 17 (3)(Spring 1990):571-97.

Carpenter, G.D., and D.J. Hahn. *The Role of Industrial Growth and Investment on Emission Trends*. Cincinnati, Ohio: Proctor and Gamble Co., 1981.

Coase, R.H. "The Problem of Social Cost." *Journal of Law and Economics* 3 (October 1960):1-44.

"Comments: Economic Efficiency in Pollution Control: EPA Issues 'Bubble' Policy for Existing Sources under Clean Air Act." *Environmental Law Reporter* 10 (January 1980):10014-18.

"Comments: EPA Approves New Jersey Generic Bubble Rule, Develops Consolidated Guidance for Controlled Trading Program." *Environmental Law Reporter* 11 (June 1981):10119-24.

"Comments: EPA's Widening Embrace of the 'Bubble' Concept: The Legality and Availability of Intra-Source Trade-Offs." *Environmental Law Reporter* 9 (February 1979):10027-31.

Connolly, Stephen J., J.H. Schwartz, E.L. Shapiro and G.M. Vogel. "Emissions Trading in Selected EPA Regions." Jellinak, Schwartz, Connolly & Freshman, Inc., 1984.

Conrad, Klaus. "Incentive Mechanisms for Environmental Protection under Asymmetric Information: A Case Study." *Applied Economics* 23 (5)(May 1991):871-80.

Cook, Brian. "The Politics of Regulatory Reform: An Analysis of Policy Choice in Environmental Regulation." Ph.D. dissertation. University of Maryland, 1984.

Courant, Carl. "Emission Reductions from Shutdowns: Their Role in Banking and Trading Systems." U.S. Environmental Protection Agency. U.S. Government Printing Office, Washington, D.C., 1980.

Crandall, Robert W. *Controlling Industrial Pollution: The Economics and Politics of Clean Air*. Washington, D.C.: The Brookings Institution, 1983.

Cristofaro, Alexander, and Joel D. Scheraga. "Policy Implications of a Comprehensive Greenhouse Gas Budget." Working paper. Office of Policy, Planning, and Evaluation, U.S. Environmental Protection Agency. U.S. Government Printing Office, Washington, D.C., 1990.

Croke, Kevin, Jay Norco and Ivan Tether. "An Economic Evaluation of Netting as an Alternative to New Source Review." Air Pollution Control Association, San Francisco, California, 1984.

Currie, David P. "Direct Federal Regulation of Stationary Sources under the Clean Air Act." *University of Pennsylvania Law Review* 128 (6)(June 1980):1389-1470.

Cuscino, Thomas, Jr., Gregory E. Muleski and Chatten Cowherd, Jr. "Iron and Steel Plant Open Source Fugitive Emission Control Evaluation." U.S. Environmental Protection Agency. Industrial Environmental Research Laboratory, Research Triangle Park, North Carolina, January 1984.

Dales, J.H. *Pollution, Property, and Prices.* Toronto: University of Toronto Press, 1968.

Dasgupta, Partha, Peter Hammond and Eric Maskin. "On Imperfect Information and Optimal Pollution Control." *Review of Economic Studies* 5 (150)(1980):857-60.

David, M., and Erhard F. Joeres. "Is a Viable Implementation of TDPs Transferable?" pp. 233-48. In *Buying a Better Environment: Cost-Effective Regulation Through Permit Trading.* Madison, Wisc.: The University of Wisconsin Press, 1983.

David, M., W. Eheart, Erhard F. Joeres and E. David. "Marketable Effluent Permits for the Control of Phosphorus Effluent in Lake Michigan." *Water Resources Research* 16 (1980):263-70.

DeLong, James V. "The Bubble Case." *Administrative Law News* 10 (1)(Fall 1984):1, 6-7.

Debusschere, Michael T. "Integrating Emissions Banking and Emission Source Review Processes." Air Pollution Control Association, San Francisco, California, 1984.

———. "Section 107 Redesignation: There Must Be a Better Way." Air Pollution Control Association, San Francisco, California, 1984.

del Calvo y Gonzalez, Jorge A. "Markets in Air: Problems and Prospects of Controlled Trading." *Harvard Environmental Law Review* 5 (1981):377-430.

Dewees, Donald N. "Instrument Choice in Environmental Policy." *Economic Inquiry* 21 (January 1983):53-71.

Dinan, Terry M. "Increasing the Demand for Old Newspapers Through Marketable Permits: Will It Work?" Association of Environmental and Resource Economists Workshop, Madison, Wisconsin, June 1990.

Domenici, Pete V. "Emissions Trading: The Subtle Heresy." *The Environmental Forum* 1 (8)(December 1982):18-24.

Doniger, David D. "The Bubble on the Cusp." *The Environmental Forum* 4 (11)(March 1986):29-34.

———. "The Dark Side of the Bubble." *The Environmental Forum* 4 (3)(July 1985):32-35.

Downing, Paul B. "Bargaining in Pollution Control." *Policy Studies Journal* 11 (4)(1983):577-86.

Drayton, William. "Thinking Ahead: Getting Smarter About Regulation." *Harvard Business Review* 59 (4)(July/August 1981):38-52.

Dudek, Daniel J. "Acid Rain Emissions Reduction Trading." U.S. Senate Subcommittee on Environmental Protection. U.S. Government Printing Office, Washington, D.C., October 4, 1989.

———. "Acid Rain: An Opportunity for Environmental Perestroika in the

United States." International Workshop, "Economic Mechanisms for Environmental Protection." Jelenia Gora, Poland, September 1989.

———. "Assessing the Implications of Changes in Carbon Dioxide Concentrations and Climate for Agriculture in the United States," pp. 428-50. In *Preparing for Climate Change*. Proceedings of the First North American Conference on Preparing for Climate Change. Rockville, Md.: Government Institutes, Inc., April 1988.

———. "Building Markets for Greenhouse Gas Management." U.S. House of Representatives, Subcommittee on Energy and Power. U.S. Government Printing Office, Washington, D.C., June 19, 1991.

———. "Carrots or Sticks: What's Best for Source Reduction?" APCA Specialty Conference, "Waste Minimization," Baltimore, Maryland, October 1988.

———. "Chlorofluorocarbon Policy: Choices and Consequences." Environmental Defense Fund, New York, New York, April 1987.

———. "Climate Change and Agriculture: Implications for Resource Management." Soil and Water Conservation Society, Columbus, Ohio, July/August 1988.

———. "Climate Change Impacts Upon Agriculture and Resources: A Case Study of California." In J.B. Smith and D.A. Tirpak, eds., *The Potential Effects of Global Climate Change on the United States*. US EPA-230-05-89-0##. Washington, D.C.

———. "Corporations in the Ecosystem: Predators or Prey?" Corporations in the Ecosystem Series. Yale University, New Haven, Connecticut, January 1991.

———. "Creating Self-Financing Environmental Markets." *Environmental Finance* 1 (4)(Winter 1991/92):511-16.

———. "Depletion of the Earth's Stratospheric Ozone Layer and the Release of Manufactured Ozone Depleting Chemicals." U.S. House of Representatives, Committee on Energy and Commerce, Subcommittee on Health and the Environment. U.S. Government Printing Office, Washington, D.C., March 9, 1987.

———. "The Ecology of Agriculture, Environment and Economy." Technical Workshop, "Developing Policies for Responding to Future Climatic Change," Villach, Austria, September/October 1987.

———. "Economic Instruments for Environmental Protection." Workshop on Carbon/Energy-Related Taxes, Environment and Taxation Working Group, Ontario Fair Tax Commission, Ontario, Canada, April 1992.

———. "Emissions Trading: Environmental Perestroika or Flimflam?" *The Electricity Journal* 2 (9)(November 1989):32-43.

———. "Energy and Environmental Policy: The Role of Markets." *Natural Resources & Environment* 6 (2)(Fall 1991):22-25, 59-61.

———. "Environmental Policy and Innovation." U.S. EPA Workshop on Alternative Control Strategies for Protecting the Ozone Layer, U.S. Government Printing Office, Washington, D.C., July 1986.

———. "From Central Planning to Markets: Restructuring Environmental

Policy." European Association of Environmental and Resource Economists, Krakow, Poland, June 1992.
———. "Global Atmospheric Change and Domestic Forest Resources." U.S. Senate, Committee on Energy and Natural Resources. U.S. Government Printing Office, Washington, D.C., September 19, 1988.
———. "Global Climate Change." U.S. Senate, Committee on Energy and Natural Resources. U.S. Government Printing Office, Washington, D.C., June 23, 1988.
———. "Harmonizing Economic and Environmental Objectives Through Trade." Malente Symposium IX, Timmerdorfer Strand, Germany, November 1991.
———. "Implications of Global Warming for Natural Resources." U.S. House of Representatives, Subcommittee on Water and Power Resources. U.S. Government Printing Office, Washington, D.C., September 27, 1988.
———. "Integrating Energy and the Environment in the Marketplace." In Olav Hohmeyer and Richard L. Ottinger, eds., *External Environmental Costs of Electric Power*. Berlin, New York: Springer-Verlag, 1991.
———. "Irrigation, Drainage, Markets and Policies: Their Current Environmental Implications." U.S. Committee on Irrigation and Drainage Regional Meeting, Washington, D.C., September 1986.
———. "Lessons from U.S. Experiments in Environmental Reform." In Z. Bochniarz and R. Bolan, eds., *Designing Institutions for Sustainable Development: A New Challenge for Poland*. Bialystok, Poland: Bialystok Technical University, June 1991.
———. "Long-term Strategy on the Environment." U.S. House of Representatives, Ways and Means Committee. U.S. Government Printing Office, Washington, D.C., March 6, 1990.
———. "Marketable Instruments for Managing Global Atmospheric Problems." Western Economics Association, Vancouver, British Columbia, Canada, July 1987.
———. "The Nexus of Agriculture, Environment and the Economy under Climate Change." In Richard E. Wyman, *Global Climate Change and Life on Earth*. New York: Routledge, Chapman and Hall Publishers, 1991.
———. "Offsetting New CO_2 Emissions." Environmental Defense Fund, New York, New York, May 1989, p. 20.
———. "Preserving Tropical Forests and Climate: The Role of Trees in Greenhouse Gas Emissions Trading." EDF/UNCTAD Workshop on Tradable Carbon Emission Entitlements, Rio de Janiero, Brazil, June 1992.
———. "Stratospheric Ozone Depletion: The Case for Policy Action." Environmental Defense Fund, New York, New York, September 1986.
———. "Stratospheric Ozone Depletion Policy: Choices and Consequences." Washington Conference on CFCs and Ozone Protection Programs. U.S. Government Printing Office, Washington, D.C., March 1987.
———. "Types and Classes of Natural Resource Economic Models." Symposium, "Do Natural Resource Economists Build Edsels?" American Agri-

cultural Economics Association, West Lafayette, Indiana, August 1983.
Dudek, Daniel J., and P. Geoffrey Allen. "Estimating Crop Yield Insurance Premium Rates." *Journal of the Northeastern Agricultural Economics Council* XIII (1)(May 1984):119-27.
Dudek, Daniel J., and Joseph Goffman. "Jump Starting the Allowance Market." *Electricity Journal* 5 (5)(June 1992):13-17.
Dudek, Daniel J., and Gerald L. Horner. "An Analytical System for the Evaluation of Land Use and Water Quality Policy Impacts upon Irrigated Agriculture," pp. 537-68. In Dan Yaron and Charles Tapiero, eds., *Operations Research in Agriculture and Water Resources*. Amsterdam, Holland: North-Holland Publishing Co., 1980.
———. "The Derived Demand for Irrigation Scheduling Services." *American Journal of Agricultural Economics* 62 (5)(December 1980):1112.
———. "External Effects of Efficient Irrigation." *American Journal of Agricultural Economics* 65 (5)(December 1983):1189.
———. "Integrated Resource and Environmental Planning." Symposium, "The Political Economy of Farmland Retention in the West," Western Agricultural Economics Association, Las Cruces, New Mexico, July 1980.
———. "Return Flow Control Policy and Income Distribution among Irrigators." *American Journal of Agricultural Economics* 63 (3)(August 1981):438-46.
———. "Section 208: An Opportunity for Resource Planning by the Book." Symposium, "Agricultural Pollution: Are We Asking the Right Questions?" American Agricultural Economics Association, Pullman, Washington, July 1979.
Dudek, Daniel J., and Richard Howitt. "Further Evidence on the Effectiveness of the California Land Conservation Act." *American Journal of Agricultural Economics* 61 (5)(December 1979):1140.
Dudek, Daniel J., and Alice M. LeBlanc. "Evaluating Firm Response: The Role of Economic Analysis in Environmental Policy Making." Applied Econometrics Association XXXIIIrd International Conference, Geneva, Switzerland, January 1992.
———. "Offsetting New CO_2 Emissions: A Rational First Greenhouse Policy Step." *Contemporary Policy Issues* (Western Economic Association International, Long Beach, California) 8 (3)(July 1990):29-41.
Dudek, Daniel J., and Gary Lynne. "Problems in Resource Model Implementation and Use, or, Do Natural Resource Economists Build Edsel?" Institute of Food and Agricultural Sciences, Gainesville, Florida, July 1983.
Dudek, Daniel J., and Robert McKusick. "A Vertically Integrated Projections System for Agricultural Land Use Planning." Western Agricultural Economics Association, Reno, Nevada, July 1975.
Dudek, Daniel J., and Michael Oppenheimer. "The Implications of Health and Environmental Effects for Policy." *Effects of Changes in Stratospheric Ozone and Global Climate* (U.S. Environmental Protection Agency and

U.N. Environment Programme) 1 (August 1986):357-79.

Dudek, Daniel J., and John Palmisano. "Emissions Trading: Why Is This Thoroughbred Hobbled?" *Columbia Journal of Environmental Law* 13 (2)(1988):217-56.

Dudek, Daniel J., and James T.B. Tripp. "Institutional Guidelines for Designing Successful Transferable Rights Programs." *Yale Journal on Regulation* 6 (2)(Summer 1989):369-91.

———. "The Swampbuster Provisions of the Food Security Act of 1985: Stronger Wetland Conservation If Properly Implemented and Enforced." *Environmental Law Reporter* 16 (May 1986): 10120-23.

Dudek, Daniel J., and Heidi Wendel. "The Design and Legality of an Innovative Approach to Nonpoint Source Control." American Water and Resources Association, Tampa, Florida, September 1989.

Dudek, Daniel J., and James Wilson. "Estimating the Benefits of Regionalizing Emergency Medical Service Provision." *Northeastern Journal of Agricultural & Resource Economics* XIV (2)(October 1985):144-53.

———. "Estimating the Impact of the Community Preferences upon the Benefits of Regionalizing EMS." Northeast Economic Development Symposium, Amherst, Mass., November 1985.

Dudek, Daniel J., Richard Adams and Bruce McCarl. "Implications of Long-Term Climate for Irrigated Agriculture." Western Agricultural Economics Association, Vancouver, British Columbia, Canada, July 1988.

Dudek, Daniel J., Gerald L. Horner and Marshall J. English. "The Derived Demand for Irrigation Scheduling Services." *Western Journal of Agricultural Economics* 6 (2)(December 1981):217-27.

Dudek, Daniel J., Gerald Horner and Robert McKusick. "An Economic Methodology for Evaluating 'Best Management Practices' in the San Joaquin Valley of California." National Conference on Management of Nitrogen in Irrigated Agriculture, Sacramento, California, May 1978.

———. "An Integrated Resource Evaluation Methodology for the Control of Irrigation Return Flows." Economics, Statistics and Cooperative Service, U.S. Department of Agriculture. U.S. Government Printing Office, Washington, D.C., March 1978.

Dudek, Daniel J., Gordon King and Harold Carter. "California Agricultural Crop Statistics: A Data Base Management Approach." Department of Agricultural Economics, University of California-Davis, Davis, California, August 1980.

———. "Projections of California Crop Production to 1985." *California Agriculture* 31 (1)(January 1978):4-6.

Dudek, Daniel J., Zbigniew Kulczynski and Tomas Zylicz. "Implementing Tradable Rights in Poland." European Association of Environmental and Resource Economists, Krakow, Poland, June 1992.

Dudek, Daniel J., Alice M. LeBlanc and Fernando Allegretti. "Disappearing Ducks: The Effect of Climate Change on North Dakota's Waterfowl." Environmental Defense Fund, New York, New York, September 1990, p.

33.
Dudek, Daniel J., Alice M. LeBlanc and Peter Miller. "SO_2 and CO_2: Consistent Policymaking in a Greenhouse." Environmental Defense Fund, New York, New York, January 1990, p. 31.
Dudek, Daniel J., Alice M. LeBlanc and Kenneth Sewall. "Business Responses to Environmental Policy: Lessons from CFC Regulation." In W. Michael Hoffman, Robert Frederick and Edward S. Petry, Jr., eds., *Business, Ethics, and the Environment: The Public Policy Debate*. Westport, Conn.: Quorum Books, 1990.
——. "Cutting the Cost of Environmental Policy: Lessons from Business Response to CFC Regulation." *AMBIO* XIX (6-7)(October 1990):324-28.
Dudek, Daniel J., Gilbert Metcalf and Cleve Willis. "Cross-Media Transfers of Hazardous Wastes." *Northeastern Journal of Agricultural & Resource Economics* XIII (2)(October 1984):203-09.
——. "The Demand for Regional Hazardous Waste Treatment, Storage and Disposal Facilities." In M.D. LaGrega and L.K. Hendrian, eds., *Toxic and Hazardous Waste*. Ann Arbor, Mich.: Ann Arbor Science Publishers, 1983.
——. "Economics of Hazardous Waste Management." University of Massachusetts. Amherst, Mass., February 1983.
——. "A Process Analysis of Electroplating Firm Response to Alternative Environmental Policies." *University of Massachusetts Experiment Station Bulletin* (May 1984).
Dudek, Daniel J., Richard B. Stewart and Jonathan B. Wiener. "Environmental Policy for Eastern Europe: Technology-Based Versus Market-Based Approaches." *Columbia Journal of Environmental Law* 17 (1)(1992):1-52.
Dudek, Daniel J., James T.B. Tripp and Michael Oppenheimer. "Protection of the Stratospheric Ozone Layer: International Science, Policy, Politics and Equity." *Environment* 29 (July/August 1987):43-45.
Dudek, Daniel J., R.M. Adams, B.A. McCarl and J.D. Glyer. "Implications of Global Climate Change for Western Agriculture." *Western Journal of Agricultural Economics* 13 (2) (December 1988):348-56.
Eads, George C., and Michael Fix. *Relief or Reform? Reagan's Regulatory Dilemma*. Washington, D.C.: The Urban Institute Press, 1984.
Elkin, Stephen L., and Brian J. Cook. "The Public Life of Economic Incentives." *Policy Studies Journal* 13 (4)(1985):797-813.
Elliott, William H., Jr. "Dupont Experience—Chambers Works 'Bubble' for Control of Volatile Organic Substances (VOS) under New Jersey Administrative Code 7:27-16.1 et seq." EPA/Armco Steel Bubble Conference, Philadelphia, Pennsylvania, December 1981.
"Emission Inventories and Air Quality Management: Selected Papers from an APCA Specialty Meeting." *Journal of Air Pollution Control Association* 32 (1982):1011-29.
Energy and Environmental Analysis, Inc. "Review of 37 Permits Approved EPA's Emissions Trading Program." Arlington, Virginia, 1986.
Evans, Jay. *Opportunities for Innovation: Administration of Sections 111(j) and*

113(d)(4) of the Clean Air Act and Industry's Development of Innovative Control Technology. Washington, D.C.: The Performance Development Institute, 1980.

Feldman, Paul S. "Regulatory Reform—Smarter Regulations." Air Pollution Control Association, San Francisco, California, 1984.

Ferrall, Brian L. "The Clean Air Act Amendments of 1990 and the Use of Market Forces to Control Sulfur Dioxide Emissions." *Harvard Journal on Legislation* 28 (1)(Winter 1991):235-52.

Fleckenstein, Leonard J. "Modeling Criteria: The Key to Major Reforms for Emissions Trades." Air Pollution Control Association, San Francisco, California, 1984.

Foskett, William H., Adrienne Jamieson and Jay Evans. "Innovation by Regulation: The Administration of Control Technology Requirements under the Clean Air Act." Experimental Technology Incentives Program, Washington, D.C., 1981.

Foster, J. David, and Malcolm C. Weiss. "Indirect Offsets: A Supplemental Program to Facilitate Interfirm Trades of Emission Reductions." Air Pollution Control Association, Philadelphia, Pennsylvania, 1981.

Friedlander, Ann F. *Approaches to Controlling Air Pollution.* Cambridge, Mass.: MIT Press, 1978.

Gabe, Jeff. "Tending Accounts at the Emissions Bank and Trust." *CBE Environmental Review* (September/October 1992):3-6.

Garelick, Barry, and J. David Foster. "Emissions Offsets as an Economic Commodity in California." Air Pollution Control Association, San Francisco, California, 1984.

Glass, Adam W. "The EPA's Bubble Concept after *Alabama Power.*" *Stanford Law Review* 32 (5)(May 1980):943-75.

Goddard, Haynes C. "Economic Incentives for Managing Household Solid Waste: Upstream Versus Downstream Policies." Conference on Research Developments for Improving Solid Waste Management, Cincinnati, Ohio, February 1991.

———. *Managing Solid Wastes: Economics, Technology and Institutions.* New York: Praeger Publishers, 1975.

Goklany, Indur. "Improving Pre-Construction Permitting for Air Pollution Sources." EPA/Regulatory Reform Staff. U.S. Government Printing Office, Washington, D.C., 1985.

Goldstein, Mark. "The Role of Emission Trading Programs in Regional Economic Development." Regional Planning Council, Baltimore, Maryland, 1983.

Gordon, H.S. "The Economic Theory of a Common Property Resource: The Fishery." *Journal of Political Economy* 62 (1954):124-42.

Greene, Kevin. "Bubble, Bubble, Spoil and Trouble." *CBE Environmental Review* (September/October 1982):7-10.

Greenwood, Jeremy, and R. Preston McAfee. "Externalities and Asymmetric Information." *The Quarterly Journal of Economics* CVI (1)(February

1991):103-21.
Hahn, Robert W. "Economic Prescriptions for Environmental Problems: How the Patient Followed the Doctor's Orders." *Journal of Economic Perspectives* 3 (2)(1989):95-114.
———. "Instrument Choice, Political Reform and Economic Welfare." *Public Choice* 67 (3) (December 1990):243-56.
———. "Market Power and Transferable Property Rights." *The Quarterly Journal of Economics* XCIX (4)(November 1984):753-65.
———. "A New Approach to the Design of Regulation in the Presence of Multiple Objectives." *Journal of Environmental Economics and Management* 17 (1989):195-211.
———. "The Political Economy of Environmental Regulation: Towards a Unifying Framework." *Public Choice* 65 (1)(April 1990):21-45.
———. "Regulation: Past, Present and Future." *Harvard Journal of Law and Public Policy* 13 (1)(Winter 1990):167-228.
———. "Trade-offs in Designing Markets with Multiple Objectives." *Journal of Environmental Economics and Management* 13 (1)(March 1986):1-12.
Hahn, Robert W., and Gordon L. Hester. "Marketable Permits: Lessons for Theory and Practice." *Ecology Law Quarterly* 16 (2)(1989):361-406.
Hahn, Robert W., and John A. Hird. "The Costs and Benefits of Regulation: Review and Synthesis." *Yale Journal on Regulation* 8 (1)(Winter 1991):233-78.
Hahn, Robert W., and Carole A. May. "The Behavior of the Allowance Market: Theory and Evidence." *The Electricity Journal* (American Enterprise Institute) (forthcoming).
Hahn, Robert W., and Gregory J. McRae. "Application of Market Mechanisms to Pollution." *Policy Studies Review* 1 (3)(1981-82):470-76.
Hahn, Robert W., and Roger G. Noll. "Barriers to Implementing Tradable Air Pollution Permits: Problems of Regulatory Interactions." *Yale Journal on Regulation* 1 (1)(1983):63-91.
———. "Designing a Market for Tradeable Emissions Permits." In W.A. Magat, ed., *Reform of Environmental Regulation*. Cambridge, Mass.: Ballinger Publishing Co., 1982.
———. "Implementing Tradeable Emission Permits," pp. 125-50. In LeRoy Graymer and Frederick Thompson, eds., *Reforming Social Regulation*. Beverly Hills, Calif: Sage Publications, 1982.
Hahn, Robert W., and Robert N. Stavins. "Incentive-Based Environmental Regulation: A New Era from an Old Idea?" *Ecology Law Quarterly* 18 (1)(1991):1-42.
Hall, Ridgway M., and Nancy Bryson. "Resource Conservation and Recovery Act." *Environment Law Handbook* (1985):61-107.
Harford, Jon D. "Firm Behavior Under Imperfectly Enforceable Pollution Standard and Taxes." *Journal of Environmental Economics and Management* 5 (1)(March 1978):26-43.
———. "Self-Reporting of Pollution and the Firm's Behavior Under Imperfectly

Enforceable Regulations." *Journal of Environmental Economics and Management* 14 (3)(September 1987):293-303.

Harrison, David, Jr., and Paul R. Portney. "Regulatory Reform in the Large and in the Small." In LeRoy Graymer and Frederick Thompson, eds., *Reforming Social Regulation.* Beverly Hills, Calif.: Sage Publications, 1982.

———. "Who Loses from Reform of Environmental Regulation?" In Wesley A. Magat edition of *Reform on Environmental Regulation.* Cambridge, Mass: Ballinger Publishing Co., 1982.

Healy, Robert G. *America's Industrial Future: An Environmental Perspective.* Washington, D.C.: The Conservation Foundation, 1982.

Helm, D., and D. Pearce. "Assessment: Economic Policy Towards the Environment." *Oxford Review of Economic Policy* 6 (1)(1990):1-16.

Henning, D., and W. Mangun. *Managing the Environmental Crisis: Incorporating Competing Values in Natural Resource Administration.* Durham, N.C.: Duke University Press, 1989.

Hess, Peter F., et al. "Experience of New Source Review in the San Francisco Bay Area." Air Pollution Control Association, San Francisco, California, 1984.

Himel, James, and Israel Patoka. "The Analysis of the Baltimore City Industrial Survey: Air Emission Banking, Trading, and Bubbling." Regional Planning Council, Baltimore, Maryland, 1982.

Hoff, Paul S. "The Current Status and Future Prospects for EPA's Bubble Policies." *Environmental Analyst* 4 (October 1983):3-8.

Hoffman, Elizabeth, and Matthew L. Spitzer. "The Coase Theorem: Some Experimental Tests." *Journal of Law and Economics* XXV (1)(April 1982):73-98.

Hoskins, David W. "Acid Rain, Emissions Trading and the Clean Air Act Amendments of 1989." *Columbia Journal of Environmental Law* 15 (2)(1990):329-57.

House, Peter, and Roger Shull. "Regulatory Reform: Politics and the Environment." Cambridge, Md.: Abt. Books; Lanham, Md.: University Press of America, 1985.

Huber, Peter. "Exorcists vs. Gatekeepers in Risk Regulation." *Regulation: AEI Journal on Government and Society* 7 (6)(November/December 1983):23-32.

ICF, Inc. "Memoranda Discussing Preliminary Analysis of Selected Topics Regarding New Source Emphasis in the Electric Utility Industry." Economic Analysis Division, U.S. Environmental Protection Agency. U.S. Government Printing Office, Washington, D.C., 1984.

Jubb, C., and B. Underhill. "Valuing the Environment: Theory, Methods and Proposed Application." *BIE Working Paper* 59 (1990).

Kelman, Steven J. "Economic Incentives and Environmental Policy: Politics, Ideology, and Philosophy." In Thomas C. Schelling, ed., *Incentives for Environmental Protection.* Cambridge, Mass: MIT Press, 1983.

———. "What Price Incentives?" *Economists and the Environment* 120 (1981).
Kneese, Allen V., and Blair T. Bower. *Managing Water Quality: Economics, Technology, Institutions.* Baltimore, Md.: John Hopkins University Press, 1968.
Kneese, Allen V., and Charles L. Schultze. *Economics and the Environment.* New York: Penguin Books, 1977.
———. *Pollution, Prices, and Public Policy.* Washington, D.C.: The Brookings Institution, 1975.
Kontnik, Lewis T. "The Evolving Emission Trading Program: Fish or Fowl?" *Environmental Analyst* 3 (August 1982):3-6.
Kostow, Lloyd P., and John Kowalczyk. "A Practical Emission Trading Program." *Journal of the Air Pollution Control Association* 33 (October 1983):982-84.
Krause, Florentin. "Energy Policy in the Greenhouse: Final Report." International Project for Sustainable Energy Paths, El Cerrito, California, 1989.
Krupnick, Alan J. "Costs of Alternative Policies for the Control of NO_2 in Baltimore." *Journal of Environmental Economics and Management* 13 (1986):189-97.
Krupnick, Alan J., Wallace E. Oates and Eric Van De Verg. "On Marketable Air-Pollution Permits: The Case for a System of Pollution Offsets." *Journal of Environmental Economics and Management* 10 (3)(September 1983):233-47.
Krupp, Frederic. "New Environmentalism Factors in Economic Needs." *The Wall Street Journal*, November 20, 1986.
Landau, Jack L. "*Chevron, U.S.A. v. NRDC*: The Supreme Court Declines to Burst EPA's Bubble Concept." *Environmental Law* 15 (2)(Winter 1985):285-322.
———. "Economic Dream or Environmental Nightmare? The Legality of the 'Bubble Concept' in Air and Water Pollution Control." *Boston College Environmental Affairs Law Review* 8 (1980):741-81.
———. "Who Owns the Air? The Emission Offset Concept and Its Implications." *Environmental Law* 9 (3)(Spring 1979):575-600.
Latin, Howard. "Ideal Versus Real Regulatory Efficiency: Implementation of Uniform Standards and 'Fine-Tuning' Regulatory Reforms." *Stanford Law Review* 37 (5)(May 1985):1267-1332.
Lave, Lester B., and Gilbert Omenn. *Clearing the Air: Reforming the Clean Air Act.* Washington, D.C.: The Brookings Institution, 1981.
Leone, Robert A., and John E. Jackson. "The Political Economy of Federal Regulatory Activity: The Case of Water-Pollution Controls." In Gary Fromm, ed., *Studies in Public Regulation.* Cambridge, Mass.: MIT Press, 1981.
Levin, Michael H. "The Clean Air Act Needs Sensible Emissions Trading." *The Environmental Forum* 4 (11)(March 1986):29, 32-33.
———. "Getting There: Implementing the 'Bubble' Policy," pp. 59-92. In Eugene Bardach and Robert Kagan, eds., *Social Regulation: Strategies for Reform.*

San Francisco, Calif.: Institute for Contemporary Studies, 1982.
———. "Status and Stopping Points: Building a Better Bubble at EPA." *Regulation* 9 (2)(March/April 1985):33-42.
———. "The Supreme Court's 'Bubble' Decision: What It Means." *EPA Journal* (September 1984):10-11.
Liroff, Richard A. *Air Pollution Offsets: Trading, Selling, and Banking.* Washington, D.C.: The Conservation Foundation, 1980.
———. "The Bubble Concept for Air Pollution Control: A Political and Administrative Perspective." Air Pollution Control Association, Philadelphia, Pennsylvania, 1981.
———. "The Bubble: Will It Float Free...or Deflate?" *The Environmental Forum* 4 (11)(March 1986):28-30.
———. "Reforming Air Pollution Regulation: The Toil and Trouble of EPA's Bubble." The Conservation Foundation, Washington, D.C., 1986.
Magat, Wesley A. *Reform of Environmental Regulation.* Cambridge, Mass.: Ballinger Publishing Co., 1982.
Majone, Giandomenico. "Choice among Policy Instruments for Pollution Control." *Policy Analysis* 2 (4)(Fall 1976):589-613.
Malik, Arun S. "Markets for Pollution Control When Firms Are Noncompliant." *Journal of Environmental Economics and Management* 18 (1990):97-106.
Maloney, Michael T., and Robert E. McCormick. "A Positive Theory of Environmental Quality Regulation." *Journal of Law and Economics* XXV (1)(April 1982):99-123.
Maloney, Michael T., and Bruce Yandle. "Bubbles and Efficiency." *Regulation* 4 (May/June 1980):49-52.
Malueg, David A. "Welfare Consequences of Emission Credit Trading Programs." *Journal of Environmental Economics and Management* 18 (1)(January 1990):66-77.
Manown, George A. "Experience of a Coke and Iron Production Facility with Non-Traditional Dust Control." Air Pollution Control Association, Atlanta, Georgia, 1983.
Marcus, Alfred A., Paul Sommers and Frederic A. Morris. "Alternative Arrangements for Cost Effective Pollution Abatements: The Need for Implementation Analysis." *Policy Studies Review* 1 (1981-82):477-83.
Meiburg, A. Stanley. *Protect and Enhance: 'Juridical Democracy' and the Prevention of Significant Deterioration of Air Quality.* New York: Garland, 1986.
Meidinger, Errol. "On Explaining the Development of 'Emissions Trading' in U.S. Air Pollution Regulation." *Law and Policy* 7 (4)(October 1985):479.
Melamed, Dennis. "Shutting Down for Credit." *The Environmental Forum* 2 (1)(May 1983):29-31.
Melnick, R. Shep. "The Clean Air Program: Options for the Future." *EPA Journal* 10 (September 1984):12-13.
———. "Deadlines, Common Sense, and Cynicism." *The Brookings Review* 2 (1)(Fall 1983):21-24.

———. "Pollution Deadlines and the Coalition for Failure." *The Public Interest* (Spring 1984):123-34.

———. *Regulation and the Courts: The Case of the Clean Air Act*. Washington, D.C.: The Brookings Institution, 1983.

Menell, Peter S. "Beyond the Throwaway Society: An Incentive Approach to Regulating Municipal Solid Waste." *Ecology Law Quarterly* 17 (4)(Winter 1990):655-739.

Miller, Susan E. "The Bubble Concept—A Feasible Emissions Reduction Alternative?" *Dayton Law Review* 9 (1983):65-80.

Montgomery, W. David. "Markets in Licenses and Efficient Pollution Control Programs." *Journal of Economic Theory* 5 (3)(December 1972):395-418.

Moore, John, Larry Parker, John Blodgett, James McMarthy and David Gushee. *Using Incentives for Environmental Protection: An Overview*. Washington, D.C.: U.S. Congressional Research Service, Library of Congress, 1989.

Muleski, Gregory E., Thomas Cuscino, Jr. and Chatten Cowherd, Jr. "Extended Evaluation of Unpaved Road Dust Suppressants in the Iron and Steel Industry." Environmental Research Laboratory, U.S. Environmental Protection Agency, Cincinnati, Ohio. U.S. Government Printing Office, Washington, D.C., April 1984.

Murray-Darling Basin Ministerial Council. "Salinity and Drainage Strategy." Murray-Darling Basin Ministerial Council, Canberra, 1989.

National Commission on Air Quality. *To Breathe Clean Air*. Washington, D.C.: U.S. Government Printing Office, March 1981.

"Note: Technology-Based Emission and Effluent Standards and the Achievement of Ambient Environmental Objectives." *Yale Law Journal* 91 (4)(March 1982):792-813.

Organisation for Economic Co-operation and Development (OECD). *Beverage Containers*. Paris: OECD, 1978.

———. "Economic Instruments for Environmental Protection." OECD, Paris, 1989.

———. *Economic Instruments in Solid Waste Management*. Paris: OECD, 1981.

———. *Environmental Policy: How to Apply Economic Instruments*. Paris: OECD, 1991.

———. "Integrating Environment and Economics." Background Paper on Resource Pricing 2. OECD, Paris, 1990.

———. *Occasional Paper on Public Management* (1)(1992).

———. *The Polluter Pays Principle*. Paris: OECD, 1975.

———. *Pollution Charges in Practice*. Paris: OECD, 1980.

Palmisano, John. "Comparing Environmental Markets with Standards." 1990. Forthcoming in *Canadian Journal of Economics*.

———. "Economic Prescriptions for Environmental Problems: Not Exactly What the Doctor Ordered." In J. Shogren, ed., *The Political Economy of Government Regulation*. Norwell, Mass.: Kluwer Academic Publishers, 1989.

———. "Emissions Trading Reforms: Successes and Failures." A paper delivered at the annual meeting of the Air Pollution Control Association, Detroit, Michigan, 1985.

———. "An Evaluation of Emissions Trading." Air Pollution Control Association, Atlanta, Georgia, 1983.

———. "An Evaluation of Incentive Based Regulatory Programs: The Performance of EPA." Rockefeller Foundation-Conservation Foundation Workshop, Warsaw, Poland, September 27, 1989.

———. "Have Markets for Trading Emission Reduction Credits Failed or Succeeded?" Regulatory Reform Staff, U.S. Environmental Protection Agency. U.S. Government Printing Office, Washington, D.C., 1982.

———. "Innovative Approaches for Revising the Clean Air Act." *Natural Resources Journal* 28 (1988):171-88.

———. "The Politics and Religion of Clean Air." *Regulation* (Winter 1990):21-30.

———. "Reshaping Environmental Policy: The Test Case of Hazardous Waste." *American Enterprise* 2 (1991):72-80.

———. "U.S. Environmental Policy: Past, Present and Future." A paper presented at the AEI Public Policy Conference, Washington, D.C., 1991. U.S. Government Printing Office.

———. "The United States' Experience with Economic Instruments in the Area of Air Pollution Control." First Joint US-USSR Conference on the Use of Economic Instruments for Environmental Protection, Sochi, USSR, June 9, 1990.

Palmisano, John, and Robert Axtell. "Reevaluating the Relationship Between Transferable Property Rights and Command-and-Control Regulation." CSIA Discussion Paper 92-04. JFK School of Government, Harvard University, Cambridge, Massachusetts, December 1991. Revised March 1992.

Palmisano, John, and Brent Haddad. "A Microeconomic Analysis of Emissions Trading and SO_2 Allowance Trading." Third Joint US-USSR Conference on the Use of Economic Incentives, Moscow, Russia, October 1991.

———. "The USSR's Experience with Economic Incentive Approaches to Pollution Control." American Economic Association Annual Meeting, New Orleans, Louisiana, January 4, 1992.

Palmisano, John, and Gordon L. Hester. "The Market for Bads: EPA's Experience with Emissions Trading." *Regulation* 3,4 (1987):48-53.

Palmisano, John, and Debora Martin. "The Use of Nontraditional Control Strategies in the Iron and Steel Industry: Air Bubbles, Water Bubbles, and Multi-media Based Control Strategies." Air Pollution Control Association, San Francisco, California, 1984.

Palmisano, John, and Roger G. Noll. "Environmental Markets in the Year 2000." *Journal of Risk and Uncertainty* 3 (1990):351-67.

"Paper Recycling: For Now, Too Much of a Good Thing." *New York Times*, September 6, 1989, p. A19.

Pearce, David, Anil Markandya and Edward B. Barbier. *Blueprint for a Green Economy*. London, Eng.: Earthscan Publications, Ltd., 1989.
PEDCo Environmental Inc. "Selective Analyses of the Prevention of Significant Deterioration (PSD) Program." National Commission on Air Quality, Washington, D.C., December 1980.
Pedersen, William F., Jr. "Pollution Accounting under the Clean Air Act." *The Environmental Forum* 3 (1)(May 1984):36-39.
———. "Why the Clean Air Act Works Badly." *University of Pennsylvania Law Review* 129 (5)(May 1981):1059-109.
Phillips, Joseph W. "The Effect of the Existing Regulatory Environment on Emerging Technology." Air Pollution Control Association, Atlanta, Georgia, 1983.
Pierce, Alan J. "Emissions Trading and Banking under the Clean Air Act after NRDC v. Gorsuch." *Syracuse Law Review* 34 (3)(Summer 1983):803-50.
Porter, Richard C. "A Social Benefit-Cost Analysis of Mandatory Deposits on Beverage Containers." *Journal of Environmental Economics and Management* 5 (4)(December 1978):351-75.
Powers, Thomas B. "Massachusetts' VOC Bubble Regulation: The Daily Cap and Issues It Raises." Air Pollution Control Association, San Francisco, California, 1984.
Project 88. "Harnessing Market Forces to Protect Our Environment: Initiatives for the New President." Project 88, Washington, D.C., 1989.
Ramil, John B., and William N. Cantrell. "Application of the 'Bubble Concept' to a Utility Coal Conversion." Air Pollution Control Association, Philadelphia, Pennsylvania, 1981.
Raufer, Roger K. "Come the Revolution: Technology and Economics in an Era of Environmental Change." U.N. Meeting on Energy and the Environment in the Development Process, Beijing, China, June 1991.
———. "Comments on 'Economic Priority in Global Warming' by A. Reteyum." Ecology and the Market, a Joint U.S.-Russian Conference, Moscow, USSR, October 1991.
———. "Comments on 'Methods of Estimation of Economic Damage from Environmental Pollution' by A.A. Gusev, et al." Joint U.S.-USSR Bilateral Conference on Economic Instruments for Environmental Protection, Sochi, USSR, June 1990.
———. "Efficiency Considerations in the Regulation of Air Pollution in the United States." Chinese Research Academy of Environmental Sciences, Beijing, April 1990.
———. "Emissions Trading for Acid Deposition Control." Society for Risk Analysis, Philadelphia, Pennsylvania, March 1990.
———. "Environmental Management." *Natural Resources Forum* 15 (2)(May 1991): 171-72. Book review of *Managing the Environmental Crisis* by Daniel H. Henning and William R. Mangun, Durham, N.C.: Duke University Press, 1989.
———. "Experience with a Market Approach to Pollution Control in the United

States." LEMIGAS R&D Center for Oil and Gas Technology, Jakarta, Indonesia, October 1987.

———. "Floating Bubbles, Instantaneous Caps, and Shoulder Trades: Issues in the Implementation of Emissions Trading by Electric Utilities under Acid Rain Controls." A paper prepared for Regulatory Innovations Staff, U.S. Environmental Protection Agency. U.S. Government Printing Office, Washington, D.C., May 1989.

———. "IPPs and the Market for Emission Allowances." Competitive Power Policy Forum, Washington, D.C., October 1990.

———. "Le Marche des 'Droits a Polluer.'" Petrole et Environnement Conference, Rueil-Malmaison, France, June 1991.

———. "Market-Based Environmental Development." *Natural Resources Forum* 16 (2)(May 1992).

———. "Market-Based Pollution Control Mechanisms for Acid Rain." IEEE Power Engineering Society, Philadelphia, Pennsylvania, January 1990.

———. "Trends in Risk Assessment for the Power Sector." United Nations (DCTD), New York, New York, October 1991.

———. "Will the Market in Emission Allowances Work?" In B.F. Hobbs, *Energy in the '90s*. Proceedings of a speciality conference sponsored by the Energy Division of the American Society of Civil Engineers, Pittsburgh, Pennsylvania, March 10-31, 1991. New York: The Society, c. 1991.

Raufer, Roger K., and M.A. Bernstein. "Impact of U.S. Environmental Regulations on Electric Utilities." Joint Business Conference of the USA/Republic of China Economic Councils, Kona, Hawaii, November 1989.

Raufer, Roger K., and R.K. Chaudry. "Emission Fees and Regulatory Efficiency." Meeting of the Air Pollution Control Association, Philadelphia, Pennsylvania, June 1981.

Raufer, Roger K., and Stephen L. Feldman. *Acid Rain and Emissions Trading: Implementing a Market Approach to Pollution Control*. Totowa, N.J.: Rowman & Littlefield, 1987.

———. "Analysis of the Potential for ERC Leasing Transactions under Various Acid Deposition Control Requirements." A paper prepared for the Regulatory Reform Staff, U.S. Environmental Protection Agency. U.S. Government Printing Office, Washington, D.C., January 1984.

———. "Emissions Trading and Acid Deposition Control: The Need for ERC Leasing." *Journal of the Air Pollution Control Association* 36 (5)(May 1986):574-80.

———. "Emissions Trading and What It May Mean for Acid Deposition Control." *Public Utilities Fortnightly* 114 (4)(August 16, 1984):17-25.

———. *Emissions Trading by Electric Utilities: I. A Literature Review Identifying Institutional and Implementation Concerns*. Report prepared for Energy Policy Division, Office of Policy Analysis, U.S. Environmental Protection Agency. Washington, D.C.: U.S. Government Printing Office, January 1983.

———. *Emissions Trading by Electric Utilities: II. Survey Results*. Report prepared for Energy Policy Division, Office of Policy Analysis, U.S. Environmental Protection Agency. Washington, D.C.: U.S. Government Printing Office, May 1983.

Raufer, Roger K., and J.E. Norco. "Air Pollution Control Through Economic Incentives: An Overview." Meeting of the Air Pollution Control Association, New Orleans, Louisiana, November 1981.

———. "Experience with an Emissions Trading Program." Energy-Sources Technology Conference and Exhibition, Houston, Texas, January 1983.

Raufer, Roger K., R.K. Chaudry and L.G. Hill. "Alternative Policies for the Control of Volatile Organic Compounds." American Institute of Chemical Engineers, New Orleans, Louisiana, November 1981.

Raufer, Roger K., S.L. Feldman and J.A. Jaksch. "The Case for Emission Reduction Credit Leasing." Air Pollution Control Association, Detroit, Michigan, 1985.

———. "Emissions Trading and Acid Deposition Control: The Need for ERC Leasing." Meeting of the Air Pollution Control Association, Detroit, Michigan, June 1985.

Raufer, Roger K., Lawrence G. Hill and Michael E. Samsa. "Emission Fees and TERA: An Evaluation of Policy Alternatives in the Twin Cities." *Journal of the Air Pollution Control Association* 31 (8)(August 1981):839-45.

Raufer, Roger K., J.A. Jaksch and E.C. Bodmer. "Tax and Financial Implications of ERC Leasing by Electric Utilities in Illinois." Meeting of the Air Pollution Control Association, Minneapolis, Minnesota, June 1986.

Raufer, Roger K., E.C. Bodmer, M.G. Willingham and R.S. Goldstein. "Tax and Financial Implications of ERC Leasing by Electric Utilities." A paper prepared for the Regulatory Reform Staff, U.S. Environmental Protection Agency. U.S. Government Printing Office, Washington, D.C., November 1985.

Raufer, Roger K., et al. *The Twin Cities/St. Cloud Regional Air Quality Study. Volume I. The State Implementation Plan Process.* Washington, D.C.: National Commission on Air Quality, November 1980.

———. *The Twin Cities/St. Cloud Regional Air Quality Study. Volume II. Alternative Development Scenarios.* Washington, D.C.: National Commission on Air Quality, November 1980.

———. *The Twin Cities/St. Cloud Regional Air Quality Study. Volume III. Alternative Measures for Emission Reduction.* Washington, D.C.: National Commission on Air Quality, November 1980.

———. *The Twin Cities/St. Cloud Regional Air Quality Study. Volume IV. Alternative Regulatory Policies.* Washington, D.C.: National Commission on Air Quality, November 1980.

Reed, Phillip D. "Comments: Court Upholds States' Relaxation of SO_2 Controls: Interstate Impacts, Sulfate Pollution Allowable." *Environmental Law Reporter* 13 (1)(January 1983):10036-42.

———. "*NRDC v. Gorsuch*: D.C. Circuit Bursts EPA's Nonattainment Area Bubble." *Environmental Law Reporter* 12 (October 1982):10089-96.

———. "When Is an Area That Is in Attainment Not an Attainment Area?" *Environmental Law Reporter* 16 (February 1986):10041.
Rehbinder, Eckard, and Rolf-Ulrich Sprenger. "The Emissions Trading Policy in the United States of America: An Evaluation of Its Advantages and Disadvantages and Analysis of Its Applicability in the Federal Republic of Germany." German Ministry of the Interior/U.S. Environmental Protection Agency, Frankfurt/Main-Munich, 1984.
"A Remedy for the Victims of Pollution Permit Markets." *Yale Law Journal* 92 (6)(May 1983):1022-40.
Rhinelander, Laurens H. "The Bubble Concept: A Pragmatic Approach to Regulation under the Clean Air Act." *Virginia Journal of Natural Resources* 1 (1981):178-228.
———. "The Proper Place for the Bubble Concept under the Clean Air Act." *Environmental Law Reporter* 13 (December 1983):10406-17.
Ritts, Leslie S. "Summary of Comments: August 31, 1983 Shutdown Notice, 48 Fed. Reg. 39579." Environmental Law Institute, Washington, D.C., 1984.
Ritts, Leslie S., Timothy R. Henderson and Alysia Watanabe. "Comparison of Selected State Emission Banking Rules." Environmental Law Institute, Washington, D.C., 1982.
———. "Comparison of Selected State Generic Comprehensive Emissions Trading Rules." Environmental Law Institute, Washington, D.C., 1982.
Roberts, Mark C., and Michael Spence. "Effluent Charges and Licenses Uncertainty." *Journal of Public Economics* 5 (3,4)(April-May 1976):193-208.
———."Some Problems of Implementing Marketable Pollution Rights Schemes: The Case of the Clean Air Act." In Wesley A. Magat, ed., *Reform of Environmental Regulation*. Cambridge, Mass.: Ballinger Publishing Co., 1982.
Romaine, Christopher P. "Bubbles or Alternative Control Strategies in Illinois." Air Pollution Control Association, San Francisco, California, 1984.
Rose-Ackerman, Susan. "Market Models for Water Pollution Control: Their Strengths and Weaknesses." *Public Policy* 25 (3)(Summer 1977):383-406.
Russell, Clifford S. "Controlled Trading of Pollution Permits." *Environmental Science and Technology* 15 (1)(January 1981):24-28.
———. "What Can We Get from Effluent Charges?" *Policy Analysis* 5 (2)(Spring 1979):155-80.
Sandler, Todd, and V. Kerry Smith. "Intertemporal and Intergenerational Pareto Efficiency." *Journal of Environmental Economics and Management* 2 (3)(February 1976):151-59.
Schelling, Thomas C. "Prices as Regulatory Instruments," pp. 1-40. *Incentives for Environmental Protection*. Cambridge, Mass.: MIT Press, 1983.
Simmons, L.L., C.L. Norton and M.J. DeBiase. "Fugitive Dust Emissions from Roads in Iron and Steel Mills: Compilation of Results and Use under

EPA's Emission Trading Policy." Symposium on Iron and Steel Pollution Abatement Technology, Pittsburgh, Pennsylvania, 1982.

Smith, Lowell, and Russell Randle. "Comment on Beyond the New Deal." *Yale Law Journal* 90 (6)(May 1981):1398-1411.

Stander, Leo. "Potential Particle Size Considerations in Developing Emission Trading Proposals." Air Pollution Control Association, Atlanta, Georgia, 1983.

Stavins, Robert N. "Alternative Renewable Resource Strategies: A Simulation of Optimal Use." *Journal of Environmental Economics and Management* 19 (2)(September 1990):143.

——. *Trading Conservation Investments for Water*. Berkeley, Calif.: Environmental Defense Fund, Inc., March 1983.

——. "Transaction Costs and the Performance of Markets for Pollution Control." JFK School of Government, Harvard University, Cambridge, Massachusetts, October 1990.

Stavins, Robert N., and Adam B. Jaffe. "Unintended Impacts of Public Investments on Private Decisions: The Depletion of Forested Wetlands." *American Economic Review* 80 (2)(June 1990):337-52.

Steiner, Bruce. "Fugitive Dust Control in Iron and Steel Plants." Ontario Industrial Waste Conference, Toronto, Ontario, 1984.

Stewart, Richard B. "Controlling Environmental Risks Through Economic Incentives." *Columbia Journal of Environmental Law* 13 (1988):153-69.

——. "Economics, Environment, and the Limits of Legal Control." *Harvard Environmental Law Review* 9 (1)(1985):1-22.

Streets, David G., et al. "A Regional New Source Bubble Policy: Its Advantages Illustrated for the State of Illinois." *Journal of the Air Pollution Control Association* 34 (1984):25-31.

Stukane, Thomas J. "EPA's Bubble Concept After *Chevron v. NRDC*: Who Is to Guard the Guards Themselves?" *Natural Resources Lawyer* 17 (4)(1985):647-82.

Sweeney, James L. "Economics of Depletable Resources: Market Forces and Intertemporal Bias." *The Review of Economic Studies* XLIV (1)(February 1977):125-41.

Teitenberg, Thomas M. "Economic Instruments for Environmental Regulation." *Oxford Review of Economic Policy* 6 (1)(1990):17-33.

——. *Emissions Trading: An Exercise in Reforming Pollution Policy*. Washington, D.C.: Resources for the Future, Inc., 1985.

——. "Transferable Discharge Permits and the Control of Stationary Source Air Pollution: A Survey and a Synthesis." *Land Economics* 56 (November 1980):391-416.

Tether, Ivan J. "Legal Issues Related to Creation, Banking and the Use of Emission Reduction Credits (ERCs); Part II: The Public Trust Doctrine." EPA 230-02-84-004. U.S. Government Printing Office, Washington, D.C., 1984.

——. "Will a Final Policy Rejuvenate the Bubble?" *The Environmental Forum*

4 (11)(March 1986):28, 31-32.
Tobin, Richard J., and John A. Jaksch. "Management Alternatives to the Clean Air Act Amendments of 1977: An Analysis of Regulatory Versus Economic Approaches." Air Pollution Control Association, Philadelphia, Pennsylvania, 1981.
Tucker, William. "Marketing Pollution: The Buying and Selling of Clean Air." *Harper's* 262 (1572)(May 1981):31-38.
U.S. Congressional Budget Office. *Carbon Charges as a Response to Global Warming: The Effects of Taxing Fossil Fuels.* Washington, D.C.: U.S. Government Printing Office, August 1990.
———. "Efficient Investments in Wastewater Treatment Plants." U.S. Government Printing Office, Washington, D.C., June 1985.
———. "Hazardous Waste Management: Recent Changes and Policy Alternatives." U.S. Government Printing Office, Washington, D.C., May 1985.
———. *Using Incentives for Environmental Protection.* Washington, D.C.: U.S. Government Printing Office, 1989.
U.S. Comptroller General. *Assessing the Feasibility of Converting Commercial Vehicle Fleets to Use Methanol as an Offset in Urban Areas.* Washington, D.C.: U.S. General Accounting Office, 1982.
———. *A Market Approach to Air Pollution Control Could Reduce Compliance Costs Without Jeopardizing Clean Air Goals.* Washington, D.C.: U.S. General Accounting Office, 1982.
U.S. Congressional Research Service. "US Primary Petrochemicals: The Superfund Taxes and Other Factors Shaping Recent Trends in Supply and Demand." Washington, D.C., August 1984.
U.S. Environmental Protection Agency (EPA). Office of Planning and Resource Management. *An Analysis of Economic Incentives to Control Emissions of Nitrogen Oxides from Stationary Sources.* Washington, D.C.: EPA, 1981.
———. "Analysis of New Source Review (NSR) Permitting Experience." Report from TRW. Washington, D.C., 1982.
U.S. Environmental Protection Agency (EPA). Regulatory Reform Staff. "Brokering Emission Reduction Credits—A Handbook." Washington, D.C., 1981.
———. *Charging Households for Waste Collection and Disposal: The Effects of Weight or Volume-Based Pricing on Solid Waste Management.* Final report. Washington, D.C.: U.S. Environmental Protection Agency, Office of Solid Waste, and Emergency Response, September 1990.
———. "Economic Incentives: Options for Environmental Protection." Office of Policy and Planning Evaluation. Washington, D.C., March 1991.
———. "Handbook for Setting User Fees." Washington, D.C., 1989.
———. "Legal Issues Related to Creation, Banking and Use of Emission Reduction Credits (ERCs)." Washington, D.C., 1982.
———. "*Report to Congress: Nonpoint Source Pollution in the United States.*" Washington, D.C.: Office of Water Programs Operations, January 1984.

Van de Verg, Eric, and Padriac Frucht. "On Trying to Be First: Maryland's Efforts to Implement EPA's Controlled Trading Policy." Western Economic Association, Annapolis, Maryland, 1981.

Victor, David G. "Tradeable Permits and Greenhouse Gas Reductions: Some Issues for U.S. Negotiators." Global Environmental Policy Project Discussion Paper G-90-06. JFK School of Government, Harvard University, Cambridge, Mass., May 1990.

Water Pollution Control Federation. "Sewer Charges for Wastewater Collection and Treatment." Washington, D.C., 1982.

Yaron, D. "A Model for the Analysis of Seasonal Aspects of Water Quality Control." *Journal of Environmental Economics and Management* 6(2)(June 1979):140-51.

Zosel, Thomas W. "Developing Effective Compliance Alternatives." Air Pollution Control Association, Atlanta, Georgia, 1983.

Zwikl, James R. "An Alternative Emission Control Plan for Industrial Plants—Birth of a Concept and Its Regulatory Implications." Air Pollution Control Association, Philadelphia, Pennsylvania, 1981.

PROJECT DIRECTORS

JOHN PALMISANO, Director

John Palmisano is director of the Environmental Policy Group for Enron Corp. From 1983 until the end of 1992, Mr. Palmisano was President of AER*X, Inc., a company specializing in air credit trading and advising both government and industry on market-based environmental reforms. Mr. Palmisano is a former manager at the U.S. Environmental Protection Agency and has worked in both government and private sector think-tanks and consulting organizations specializing in the development of market-based environmental and energy policies.

CAROLE NEVES, Deputy Project Director

Carole Neves is a project director at the National Academy of Public Administration (NAPA), where she has directed projects for a number of U.S. agencies, among them the National Aeronautics and Space Administration, Environmental Protection Agency, Small Business Administration and Federal Aviation Administration. Ms. Neves has also worked on international projects for private sector organizations, including non-profit groups. Recently, she worked on strengthening governmental capacity in Eastern Europe and the Baltics.

BIOGRAPHICAL INFORMATION ON PANEL

RICHARD A. WEGMAN, Chair. Partner, Garvey, Schubert, & Barer. Former Chief Counsel and Staff Director, Senate Committee on Governmental Affairs; Executive Director, President's Commission on a National Agenda for the Eighties.

ALVIN L. ALM. Director and Senior Vice President, Science Applications International Corporation. Former President, Alliance Technologies Corporation; Chairman of the Board, Thermo Analytical, Inc.; Deputy Administrator, U.S. Environmental Protection Agency; Assistant Secretary, Policy and Evaluation, U.S. Department of Energy.

DOUGLAS M. COSTLE. Former Dean, Vermont Law School; Counsel to law firms in Washington, D.C. and Hartford, Connecticut; Visiting Scholar, Harvard School of Public Health; Adjunct Lecturer, JFK School of Government, Harvard University; Administrator, U.S. Environmental Protection Agency; Chairman, U.S. Regulatory Council; Chairman, U.S./U.S.S.R. Joint Committee on Cooperation in the Field of Environmental Protection; and Co-chairman, U.S./People's Republic of China Environmental Protection Protocol.

J. CLARENCE DAVIES. Director, Center for Risk Management, Resources for the Future. Former Executive Director, National Commission on the Environment; Assistant Administrator for Policy, Planning and Evaluation, U.S. Environmental Protection Agency; Executive Vice President, The Conservation Foundation; Fellow, Resources for the Future; Senior Staff Member, Council on Environmental Quality, Executive Office of the President; and Assistant Professor of Politics and Public Affairs, Princeton University.

WILLIAM DRAYTON, JR. Chairman, Ashoka Society. Former Assistant Administrator, U.S. Environmental Protection Agency; Member, White House Domestic Policy Staff; Faculty, JFK School of Government, Harvard University; Faculty, Stanford Law School; and Counsel, McKinsey and Company, Inc.

ANTHONY S. EARL. Partner, Quarles & Brady. Former Governor, State of Wisconsin; Secretary, Department of Natural Resources; Secretary, Department of Administration; Member, Wisconsin State Legislature; City Attorney, Wausau, Wisconsin; and Assistant District Attorney, Marathon County, Wisconsin.

FRANK B. FRIEDMAN. Senior Vice President, Health, Environment, and Safety, Elf Atochem North America, Inc. Formerly Partner, McClintock, Weston, Benshoof, Rochefort, Rubalcava & MacCuish; Vice President, Health, Environment and Safety; Special Counsel, Occidental Petroleum Corporation; various legal and management positions with the Atlantic Richfield Company;

and Trial Attorney, Environment and Natural Resources Division, U.S. Department of Justice.

ROBERT W. HAHN. Resident Scholar, American Enterprise Institute and Adjunct Research Fellow, JFK School of Government, Harvard University. Cochairman, U.S. Alternative Fuels Council; Senior Staff Member, President's Council of Economic Advisors; and Associate Professor of Economics and Public Policy, Carnegie Mellon University.

DAVID G. HAWKINS. Senior Attorney, Natural Resources Defense Council. Former Assistant Administrator for Air, Noise, and Radiation, U.S. Environmental Protection Agency; Air Pollution Specialist, Natural Resources Defense Council; and Environmental Specialist, Stern Community Law Firm.

JAMES M. LENTS. Executive Officer, South Coast Air Quality Management District (Southern California). Former Head, Air Pollution Control Division, Colorado Department of Health; and Technical Director, Chattanooga-Hamilton County Air Pollution Control Bureau (Tennessee).

EDWARD G. SANDERS. President, Sanders International, Inc. Former President, International Planning and Analysis Center, Inc.; Staff Director, Senate Foreign Relations Committee; and Associate Director for National Security and International Affairs, Office of Management and Budget.

RICHARD E. STEWART. Professor, New York University School of Law. Former Chairman, Stewart Economics, Inc.; Senior Vice President and Chief Financial Officer, The Chubb Corporation; Senior Vice President and General Counsel, First National City Bank and First National City Corporation; Superintendent of Insurance of New York State; and President, National Association of Insurance Commissioners.

VICTORIA JEAN TSCHINKEL. Senior Consultant for Environmental Issues, Landers & Parsons. Former Secretary of the Florida Department of Environmental Regulation; and several positions at the Florida Department of Environmental Regulation and the Tall Timbers Research Station.